SOLID MECHANICS

This is a textbook for courses in departments of Mechanical, Civil and Aeronautical Engineering commonly called strength of materials or mechanics of materials. The intent of this book is to provide a background in the mechanics of solids for students of mechanical engineering while limiting the information on why materials behave as they do. It is assumed that the students have already had courses covering materials science and basic statics. Much of the material is drawn from another book by the author, *Mechanical Behavior of Materials*. To make the text suitable for Mechanical Engineers, the chapters on slip, dislocations, twinning, residual stresses, and hardening mechanisms have been eliminated and the treatments in other chapters about ductility, viscoelasticity, creep, ceramics, and polymers have been simplified.

William Hosford is a Professor Emeritus of Materials Science at the University of Michigan. He is the author of numerous research and publications books, including *Materials for Engineers; Metal Forming* third edition (with Robert M. Caddell); *Materials Science: An Intermediate Text; Reporting Results* (with David C. Van Aken); *Mechanics of Crystals and Textured Polycrystals; Mechanical Metallurgy*; and *Wilderness Canoe Tripping*.

Solid Mechanics

William Hosford

University of Michigan, Emeritus

CAMBRIDGE UNIVERSITY PRESS
Cambridge, New York, Melbourne, Madrid, Cape Town,
Singapore, São Paulo, Delhi, Mexico City

Cambridge University Press
32 Avenue of the Americas, New York NY 10013-2473, USA

Published in the United States of America by Cambridge University Press, New York

www.cambridge.org
Information on this title: www.cambridge.org/9781107632943

First published 2010
First paperback edition 2013

A catalogue record for this publication is available from the British Library

Library of Congress Cataloguing in Publication Data
Hosford, William F.
Solid Mechanics / William Hosford.
    p.   cm.
Includes bibliographical references and index.
ISBN 978-0-521-19229-3 (hardback)
1. Strength of materials.   2. Mechanical engineering – Materials.
I. Title.
TA405.H595   2010
620.1′05–dc22   2009045587

ISBN 978-0-521-19229-3 Hardback
ISBN 978-1-107-63294-3 Paperback

Contents

Preface

The intent of this book is to provide a background in the mechanics of solids for students of mechanical engineering without confusing them with too much detail on why materials behave as they do. The topics of this book are similar to those in *Deformation and Fracture of Solids* by R. M. Caddell. Much of the material is drawn from another book by the author, *Mechanical Behavior of Materials*. To make the text suitable for Mechanical Engineers, the chapters on slip, dislocations, twinning, residual stresses, and hardening mechanisms have been eliminated and the treatments in other chapters about ductility, viscoelasticity, creep, ceramics, and polymers have been simplified. If there is insufficient time or interest, the last two chapters, "Mechanical Working" and "Anisotropy," may be omitted. It is assumed that the students have already had courses covering materials science and basic statics.

I want to thank Professor Robert Caddell for the inspiration to write texts. Discussions with Professor Jwo Pan about what to include were helpful.

Conversions

To convert from	To	Multiply by
inch, in.	meter, m	0.0254
pound force, lb_f	newton, N	0.3048
pounds/inch^2	pascal, Pa	6.895×10^3
kilopound/inch^2	megapascal, MPa	6.895×10^3
kilograms/mm^2	pascals	9.807×10^6
horsepower	watts, W	7.457×10^2
horsepower	ft-lb/min	33×10^3
foot-pound	joule, J	1.356
calorie	joule, J	4.187

SI Prefixes

tera	T	10^{12}	pico	p	10^{-12}
giga	G	10^9	nano	n	10^{-9}
mega	M	10^6	micro	μ	10^{-6}
kilo	k	10^3	milli	m	10^{-3}

1 Stress and Strain

Introduction

This book is concerned with the mechanical behavior of materials. The term *mechanical behavior* refers to the response of materials to forces. Under load, materials may either deform or break. The factors that govern a material's resistance to deforming are very different than those governing its resistance to fracture. The word *strength* may refer either to the stress required to deform a material or to the stress required to cause fracture; therefore, care must be used with the term *strength*.

When a material deforms under a small stress, the deformation may be *elastic*. In this case when the stress is removed, the material will revert to its original shape. Most of the elastic deformation will recover immediately. However, there may be some time-dependent shape recovery. This time-dependent elastic behavior is called *anelasticity* or *viscoelasticity*.

A larger stress may cause *plastic* deformation. After a material undergoes plastic deformation, it will not revert to its original shape when the stress is removed. Usually, a high resistance to deformation is desirable so that a part will maintain its shape in service when stressed. On the other hand, it is desirable to have materials deform easily when forming them into useful parts by rolling, extrusion, and so on. Plastic deformation usually occurs as soon as the stress is applied. At high temperatures, however, time-dependent plastic deformation called *creep* may occur.

Fracture is the breaking of a material into two or more pieces. If fracture occurs before much plastic deformation occurs, we say the material is *brittle*. In contrast, if there has been extensive plastic deformation preceding fracture, the material is considered *ductile*. Fracture usually occurs as soon as a critical fracture stress has been reached; however, repeated applications of a somewhat lower stress may cause fracture. This is called *fatigue*.

The amount of deformation that a material undergoes is described by *strain*. The forces acting on a body are described by *stress*. Although the reader

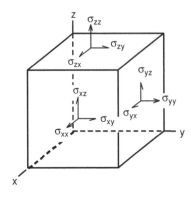

Figure 1.1. The nine components of stress acting on an infinitesimal element. The normal stress components are σ_{xx}, σ_{yy}, and σ_{zz}. The shear stress components are $\sigma_{yz}, \sigma_{zx}, \sigma_{xy}, \sigma_{zy}, \sigma_{xz}$, and σ_{yx}.

should already be familiar with these terms, they will be reviewed in this chapter.

Stress

Stress, σ, is defined as the intensity of force at a point,

$$\sigma = \partial F/\partial A \text{ as } \partial A \rightarrow 0. \tag{1.1a}$$

If the state of stress is the same everywhere in a body,

$$\sigma = F/A. \tag{1.1b}$$

A *normal stress* (compressive or tensile) is one in which the force is normal to the area on which it acts. With a *shear stress*, the force is parallel to the area on which it acts.

Two subscripts are required to define a stress. The first subscript denotes the normal to the plane on which the force acts, and the second subscript identifies the direction of the force.* For example, a tensile stress in the x-direction is denoted by σ_{xx}, indicating that the force is in the x-direction and it acts on a plane normal to x. For a shear stress, σ_{xy}, a force in the y-direction acts on a plane normal to x.

Because stresses involve both forces and areas, they are tensor rather than vector quantities. Nine components of stress are needed to describe fully a state of stress at a point, as shown in Figure 1.1. The stress component $\sigma_{yy} = F_y/A_y$ describes the tensile stress in the y-direction. The stress component $\sigma_{zy} = F_y/A_z$ is the shear stress caused by a shear force in the y direction acting on a plane normal to z.

Repeated subscripts denote normal stresses (e.g. $\sigma_{xx}, \sigma_{yy}, \ldots$), whereas mixed subscripts denote shear stresses (e.g. $\sigma_{xy}, \sigma_{zx} \ldots$) . In *tensor* notation,

---

* Use of the opposite convention should cause no confusion as $\sigma_{ij} = \sigma_{ji}$.

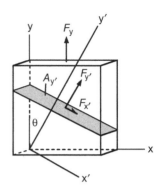

Figure 1.2. Stresses acting on an area, A', under a normal force, F_y. The normal stress is $\sigma_{y'y} = F_y/A_{y'} = F_y\cos\theta/(A_y/\cos\theta) = \sigma_{yy}\cos^2\theta$. The shear stress is $\tau_{y'x'} = F_x/A_{y'} = F_y\sin\theta/(A_{yx}/\cos\theta) = \sigma_{yy}\cos\theta\sin\theta$.

the state of stress is expressed as

$$\sigma_{ij} = \begin{vmatrix} \sigma_{xx} & \sigma_{xy} & \sigma_{xz} \\ \sigma_{yx} & \sigma_{yy} & \sigma_{yz} \\ \sigma_{zx} & \sigma_{zy} & \sigma_{zz} \end{vmatrix}, \tag{1.2}$$

where i and j are iterated over x, y, and z. Except where tensor notation is required, it is often simpler to use a single subscript for a normal stress and to denote a shear stress by τ,

$$\sigma_x = \sigma_{xx}, \quad \text{and} \quad \tau_{xy} = \sigma_{xy}. \tag{1.3}$$

A stress component, expressed along one set of axes, may be expressed along another set of axes. Consider the case in Figure 1.2. The body is subjected to a stress $\sigma_{yy} = F_y/A_y$. It is possible to calculate the stress acting on a plane whose normal, y', is at an angle θ to y. The normal force acting on the plane is $F_{y'} = F_y\cos\theta$ and the area normal to y' is $A_y/\cos\theta$, so

$$\sigma_{y'} = \sigma_{y'y'} = F_{y'}/A_{y'} = (F_y\cos\theta)/(A_y/\cos\theta) = \sigma_y\cos^2\theta. \tag{1.4a}$$

Similarly, the shear stress on this plane acting in the x' direction, $\tau_{y'x'}(=\sigma_{y'x'})$, is given by

$$\tau_{y'x'} = \sigma_{y'x'} = F_{x'}/A_{y'} = (F_y\sin\theta)/(A_y/\cos\theta) = \sigma_y\cos\theta\sin\theta. \tag{1.4b}$$

Note that the transformation equations involve the product of two cosine and/or sine terms.

Sign Convention

When we write $\sigma_{ij} = F_i/A_j$, the term σ_{ij} is positive if i and j are either both positive or both negative. On the other hand, the stress component is negative for a combination of i and j in which one is positive and the other is negative. For example, in Figure 1.3 the term σ_{xx} is positive on both sides of the element because both the force and normal to the area are negative on the left and positive on the right. The stress τ_{yx} is negative because on the top surface y is

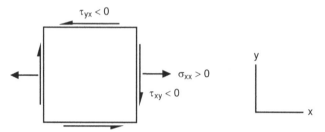

Figure 1.3. The normal stress, σ_{xx}, is positive because the direction of the force F_x and the plane A_x are either both positive (right) or both negative (left). The shear stress, τ_{xy} and τ_{yx}, are negative because the direction of the force and the normal to the plane have opposite signs.

positive and the x-direction force is negative, and on the bottom surface the x-direction force is positive and the normal to the area, y is negative. Similarly, τ_{xy} is negative.

Pairs of shear stress terms with reversed subscripts are always equal. A moment balance requires that $\tau_{ij} = \tau_{ji}$. If they were not, the element would rotate (Figure 1.4). For example, $\tau_{yx} = \tau_{xy}$. Therefore, we can write in general that $\Sigma M_A = \tau_{yx} = \tau_{xy} = 0$, so

$$\sigma_{ij} = \sigma_{ji}, \quad \text{or} \quad \tau_{ij} = \tau_{ji}. \tag{1.5}$$

This makes the stress tensor matrix symmetric about the diagonal.

Transformation of Axes

Frequently, it is useful to change the axis system on which a stress state is expressed. For example, we may want to find the shear stress on a inclined plane from the external stresses. Another example is finding the normal stress across a glued joint in a tube subjected to tension and torsion. In general, a stress state expressed along one set of orthogonal axes (e.g., m, n, and p) may be expressed along a different set of orthogonal axes (e.g., i, j, and k).

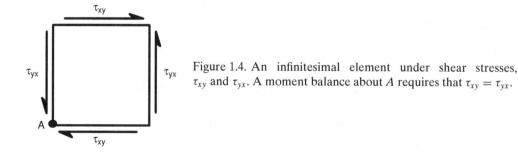

Figure 1.4. An infinitesimal element under shear stresses, τ_{xy} and τ_{yx}. A moment balance about A requires that $\tau_{xy} = \tau_{yx}$.

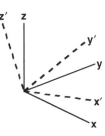

Figure 1.5. Two orthogonal coordinate systems, (x, y, z) and (x', y', z').
The stress state may be expressed in terms of either.

The general form of the transformation is

$$\sigma_{ij} = \sum_{n=1}^{3} \sum_{m=1}^{3} \ell_{im}\ell_{jn}\sigma_{mn}. \tag{1.6}$$

The term ℓ_{im} is the cosine of the angle between the i and m axes, and ℓ_{jn} is
the cosine of the angle between the j and n axes. The summations are over the
three possible values of m and n, namely m, n, and p. This is often written as

$$\sigma_{ij} = \ell_{im}\ell_{jn}\sigma_{mn} \tag{1.7}$$

with the summation implied. The stresses in the (x, y, z) coordinate system in
Figure 1.5 may be transformed onto the (x', y', z') coordinate system by

$$\begin{aligned}
\sigma_{x'x'} &= \ell_{x'x}\ell_{x'x}\sigma_{xx} + \ell_{x'y}\ell_{x'x}\sigma_{yx} + \ell_{x'z}\ell_{x'x}\sigma_{zx} \\
&\quad + \ell_{x'x}\ell_{x'y}\sigma_{xy} + \ell_{x'y}\ell_{x'y}\sigma_{yy} + \ell_{x'z}\ell_{x'y}\sigma_{zy} \\
&\quad + \ell_{x'x}\ell_{x'z}\sigma_{xz} + \ell_{x'y}\ell_{x'z}\sigma_{yz} + \ell_{x'z}\ell_{x'z}\sigma_{zz}
\end{aligned} \tag{1.8a}$$

and

$$\begin{aligned}
\sigma_{y'y'} &= \ell_{x'x}\ell_{y'x}\sigma_{xx} + \ell_{x'y}\ell_{y'x}\sigma_{yx} + \ell_{x'z}\ell_{y'x}\sigma_{zx} \\
&\quad + \ell_{x'x}\ell_{y'y}\sigma_{xy} + \ell_{x'y}\ell_{y'y}\sigma_{yy} + \ell_{x'z}\ell_{y'y}\sigma_{zy} \\
&\quad + \ell_{x'x}\ell_{y'z}\sigma_{xz} + \ell_{x'y}\ell_{y'z}\sigma_{yz} + \ell_{x'z}\ell_{y'z}\sigma_{zz}.
\end{aligned} \tag{1.8b}$$

These equations may be simplified with the notation in Equation 1.3 using
Equation 1.5,

$$\begin{aligned}
\sigma_{x'} &= \ell_{x'x}^2\sigma_x + \ell_{x'y}^2\sigma_y + \ell_{x'z}^2\sigma_z \\
&\quad + 2\ell_{x'y}\ell_{x'z}\tau_{yz} + 2\ell_{x'z}\ell_{x'x}\tau_{zx} + 2\ell_{x'x}\ell_{x'y}\tau_{xy}
\end{aligned} \tag{1.9a}$$

and

$$\begin{aligned}
\tau_{x'z'} &= \ell_{x'x}\ell_{y'x}\sigma_{xx} + \ell_{x'y}\ell_{y'y}\sigma_{yy} + \ell_{x'z}\ell_{y'z}\sigma_{zz} \\
&\quad + (\ell_{x'y}\ell1_{y'z} + \ell_{x'z}\ell_{y'y})\tau_{yz} + (\ell_{x'z}\ell_{y'x} + \ell_{x'x}\ell_{y'z})\tau_{zx} \\
&\quad + (\ell_{x'x}\ell_{y'y} + \ell_{x'y}\ell_{y'x})\tau_{xy}.
\end{aligned} \tag{1.9b}$$

Now reconsider the transformation in Figure 1.2. Using equations 1.9a and 19b
with σ_{yy} as the only finite term on the (x, y, z) axis system,

$$\sigma_{y'} = \ell_{y'y}^2\sigma_{yy} = \sigma_y\cos^2\theta \quad \text{and} \quad \tau_{x'y'} = \ell_{x'y}\ell_{y'y}\sigma_{yy} = \sigma_y\cos\theta\sin\theta \tag{1.10}$$

in agreement with Equations 1.4a and 1.4b.

Principal Stresses

It is always possible to find a set of axes $(1, 2, 3)$ along which the shear stress components vanish. In this case the normal stresses, σ_1, σ_2, and σ_3, are called *principal stresses* and the 1, 2, and 3 axes are the *principal stress axes*. The magnitudes of the principal stresses, σ_p, are the three roots of

$$\sigma_p^3 - I_1\sigma_p^2 - I_2\sigma_p - I_3 = 0, \tag{1.11}$$

where

$$
\begin{aligned}
I_1 &= \sigma_{xx} + \sigma_{yy} + \sigma_{zz}, \\
I_2 &= \sigma_{yz}^2 + \sigma_{zx}^2 + \sigma_{xy}^2 - \sigma_{yy}\sigma_{zz} - \sigma_{zz}\sigma_{xx} - \sigma_{xx}\sigma_{yy} \\
I_3 &= \sigma_{xx}\sigma_{yy}\sigma_{zz} + 2\sigma_{yz}\sigma_{zx}\sigma_{xy} - \sigma_{xx}\sigma_{yz}^2 - \sigma_{yy}\sigma_{zx}^2 - \sigma_{zz}\sigma_{xy}^2.
\end{aligned}
\tag{1.12}
$$

The first invariant is $I_1 = -p/3$, where p is the pressure. I_1, I_2, and I_3 are independent of the orientation of the axes and are therefore called *stress invariants*. In terms of the principal stresses, the invariants are

$$
\begin{aligned}
I_1 &= \sigma_1 + \sigma_2 + \sigma_3 \\
I_2 &= -\sigma_{22}\sigma_{33} - \sigma_{33}\sigma_{11} - \sigma_{11}\sigma_{22} \\
I_3 &= \sigma_{11}\sigma_{22}\sigma_{33}.
\end{aligned}
\tag{1.13}
$$

EXAMPLE PROBLEM #1.1: Find the principal stresses in a body under the stress state, $\sigma_x = 10, \sigma_y = 8, \sigma_z = -5, \tau_{yz} = \tau_{zy} = 5, \tau_{zx} = \tau_{xz} = -4$, and $\tau_{xy} = \tau_{yx} = -8$, where all stresses are in MPa.

Solution: Using Equation 1.13, $I_1 = 10 + 8 - 5 = 13, I_2 = 5^2 + (-4)^2 + (-8)^2 - 8(-5) - (-5)10 - 10.8 = 115, I_3 = 10.8(-5) + 2.5(-4)(-8) - 10.5^2 - 8(-4)^2 - (-5)(-8)^2 = -138$ MPa.

Solving Equation 1.11, $\sigma_p^3 - 13\sigma_p^2 - 115\sigma_p + 138 = 0, \sigma_p = 1.079, 18.72, -6.82$ MPa.

Mohr's Stress Circles

In the special case where there are no shear stresses acting on one of the reference planes (e.g., $\tau_{zy} = \tau_{zx} = 0$), the normal to that plane, z, is a principal stress direction and the other two principal stress directions lie in the plane. This is illustrated in Figure 1.6. For these conditions, $\ell_{x'z} = \ell_{y'z} = 0, \tau_{zy} = \tau_{zx} = 0, \ell_{x'x} = \ell_{y'y} = \cos\phi$, and $\ell_{x'y} = -\ell_{y'x} = \sin\phi$. The variation of the shear stress component $\tau_{x'y'}$ can be found by substituting these conditions into the stress transformation Equation (1.8b). Substituting $\ell_{x'z} = \ell_{y'z} = 0$,

$$\tau_{x'y'} = \cos\phi\sin\phi(-\sigma_{xx} + \sigma_{yy}) + (\cos^2\phi - \sin^2\phi)\tau_{xy}. \tag{1.14a}$$

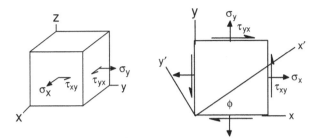

Figure 1.6. Stress state to which Mohr's circle treatment applies. Two shear stresses, τ_{yz} and τ_{zx}, are zero.

Similar substitution into the expressions for $\sigma_{x'}$ and $\sigma_{y'}$ results in

$$\sigma_{x'} = \cos^2\phi\,\sigma_x + \sin^2\phi\,\sigma_y + 2\cos\phi\sin\phi\,\tau_{xy} \tag{1.14b}$$

and

$$\sigma_{y'} = \sin^2\phi\,\sigma_x + \cos^2\phi\,\sigma_y + 2\cos\phi\sin\phi\,\tau_{xy}. \tag{1.14c}$$

These can be simplified by substituting the trigonometric identities $\sin 2\phi = 2\sin\phi\cos\phi$ and $\cos 2\phi = \cos^2\phi - \sin\phi$,

$$\tau_{x'y'} = -[(\sigma_x - \sigma_y)/2]\sin2\phi + \tau_{xy}\cos2\phi \tag{1.15a}$$
$$\sigma_{x'} = (\sigma_x + \sigma_y)/2 + [\sigma_x - \sigma_y)/2]\cos2\phi\,\tau_{xy}\sin2\phi \tag{1.15b}$$

and

$$\sigma_{y'} = (\sigma_x + \sigma_y)/2 - [(\sigma_x - \sigma_y)/2]\cos2\phi + \tau_{xy}\sin2\phi. \tag{1.15c}$$

Setting $\tau_{x'y'} = 0$ in Equation 1.15a, ϕ becomes the angle, θ, between the principal stresses axes and the x and y axes (see Figure 1.7):

$$\tau_{x'y'} = 0 = \sin2\theta(\sigma_x - \sigma_y)/2 + \cos2\theta\,\tau_{xy} \quad \text{or} \quad \tan 2\theta = \tau_{xy}/[(\sigma_x - \sigma_y)/2].$$
$$\tag{1.16}$$

The principal stresses σ_1 and σ_2 are the values of $\sigma_{x'}$ and $\sigma_{y'}$ for this value of ϕ,

$$\sigma_{1,2} = (\sigma_x + \sigma_y)/2 \pm [\sigma_x - \sigma_y)/2]\cos2\theta + \tau_{xy}\sin2\theta \quad \text{or}$$
$$\sigma_{1,2} = (\sigma_x + \sigma_y)/2 \pm (1/2)[(\sigma_x - \sigma_y)^2 + 4\tau_{xy}^2]^{1/2} \tag{1.17}$$

A Mohr's circle diagram is a graphical representation of Equations 1.16 and 1.17. It plots as a circle with a radius $(\sigma_1 - \sigma_2)/2$ centered at

$$(\sigma_1 + \sigma_2)/2 = (\sigma_x + \sigma_y)/2 \tag{1.17a}$$

as shown in Figure 1.7. The normal stress components, σ, are represented on the ordinate and the shear stress components, τ, on the abscissa. Consider the

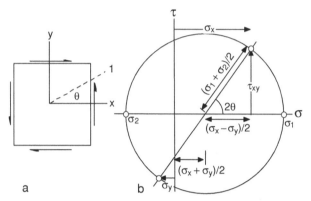

Figure 1.7. Mohr's circles for stresses showing the stresses in the x-y plane. Note that the 1-axis is rotated clockwise from the x-axis in real space (a), whereas in the Mohr's circle diagram (b) the 1-axis is rotated counterclockwise from the x-axis.

triangle in Figure 1.7b. Using the Pythagorean theorem, the hypotenuse is

$$(\sigma_1 - \sigma_2)/2 = \left\{[(\sigma_x + \sigma_y)/2]^2 + \tau_{xy}^2\right\}^{1/2} \tag{1.17b}$$

and

$$\tan(2\theta) = [\tau_{xy}/[(\sigma_x + \sigma_y)/2]. \tag{1.17c}$$

The full three-dimensional stress state may be represented by three Mohr's circles (Figure 1.7).

The three principal stresses, σ_1, σ_2, and σ_3, are plotted on the horizontal axis. The circles connecting these represent the stresses in the 1–2, 2–3 and 1–3 planes. The largest shear stress may be either $(\sigma_1 - \sigma_2)/2$, $(\sigma_2 - \sigma_3)/2$ or $(9\sigma_1 - \sigma_3)/2$.

> **EXAMPLE PROBLEM #1.2:** A body is loaded under stresses $\sigma_x = 150\,\text{MPa}$, $\sigma_y = 60\,\text{MPa}$, $\tau_{xy} = 20\,\text{MPa}$, $\sigma_z = \tau_{yz} = \tau_{zx} = 0$. Find the three principal stresses, sketch the three-dimensional Mohr's circle diagram for this stress state, and find the largest shear stress in the body.

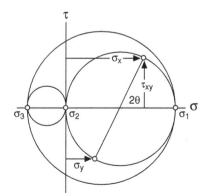

Figure 1.8. Three Mohr's circles representing a stress state in three dimensions. The three circles represent the stress states in the 2–3, 3–1, and 1–2 planes.

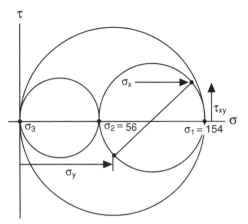

Figure 1.9. Mohr's circles for Example #1.2.

Solution: $\sigma_1, \sigma_2 = (\sigma_x + \sigma_y)/2 \pm \{[(\sigma_x - \sigma_y)/2]^2 + \tau_{xy}^2\}^{1/2} = 154.2, 55.8\,\text{MPa}$, $\sigma_3 = \sigma_z = 0$. Figure 1.9 is the Mohr's circle diagram. Note that the largest shear stress, $\tau_{\max} = (\sigma_1 - \sigma_3)/2 = 77.1\,\text{MPa}$, is not in the 1–2 plane.

Strains

An infinitesimal normal strain is defined by the change of length, L, of a line:

$$d\varepsilon = dL/L. \tag{1.18}$$

Integrating from the initial length, L_o, to the current length, L,

$$\varepsilon = \int dL/L = \ln(L/L_o) \tag{1.19}$$

This finite form is called *true strain* (or *natural strain, logarithmic strain*). Alternatively, *engineering* or *nominal strain, e,* is defined as

$$e = \Delta L/L_o. \tag{1.20}$$

If the strains are small, then the engineering and true strains are nearly equal. Expressing $\varepsilon = \ln(L/L_o) = \ln(1 + e)$ as a series expansion, $\varepsilon = e - e^2/2 + e^3/3! \ldots \ldots$ so as $e \to 0$, $\varepsilon \to e$. This is illustrated in the following example.

EXAMPLE PROBLEM #1.3: Calculate the ratio of e/ε for several values of e.

Solution: $e/\varepsilon = e/\ln(1 + e)$. Evaluating:

for $e = 0.001$, $e/\varepsilon = 1.0005$;
for $e = 0.01$, $e/\varepsilon = 1.005$;
for $e = 0.02$, $e/\varepsilon = 1.010$;
for $e = 0.05$, $e/\varepsilon = 1.025$;
for $e = 0.10$, $e/\varepsilon = 1.049$;
for $e = 0.20$, $e/\varepsilon = 1.097$;
for $e = 0.50$, $e/\varepsilon = 1.233$.

Note that the difference e and ε between is less than 1% for $e < 0.02$

There are several reasons that true strains are more convenient than engineering strains:

1. True strains for equivalent amounts of deformation in tension and compression are equal except for sign.
2. True strains are additive. For a deformation consisting of several steps, the overall strain is the sum of the strains in each step.
3. The volume change is related to the sum of the three normal strains. For constant volume, $\varepsilon_x + \varepsilon_y + \varepsilon_z = 0$.

These statements are not true for engineering strains, as illustrated in the following examples.

EXAMPLE PROBLEM #1.4: An element 1 cm long is extended to twice its initial length (2 cm) and then compressed to its initial length (1 cm).
a. Find the true strains for the extension and compression.
b. Find the engineering strains for the extension and compression.

Solution: a. During the extension, $\varepsilon = \ln(L/L_o) = \ln 2 = 0.693$, and during the compression $\varepsilon = \ln(L/L_o) = \ln(1/2) = -0.693$.

b. During the extension, $e = \Delta L/L_o = 1/1 = 1.0$, and during the compression $e = \Delta L/L_o = -1/2 = -0.5$.

Note that with engineering strains, the magnitude of strain to reverse the shape change is different.

EXAMPLE PROBLEM #1.5: A bar 10 cm long is elongated by (1) drawing to 15 cm, and then (2) drawing to 20 cm.

a) Calculate the engineering strains for the two steps and compare the sum of these with the engineering strain calculated for the overall deformation.
b) Repeat the calculation with true strains.

Solution: (a) For step 1, $e_1 = 5/10 = 0.5$; for step 2, $e_2 = 5/15 = 0.333$. The sum of these is 0.833, which is less than the overall strain, $e_{tot} = 10/10 = 1.00$

(b) For step 1, $\varepsilon_1 = \ln(15/10) = 0.4055$; for step 2, $\varepsilon_1 = \ln(20/15) = 0.2877$. The sum is 0.6931, and the overall strain is $\varepsilon_{tot} = \ln(15/10) + \ln(20/15) = \ln(20/10) = 0.6931$.

EXAMPLE PROBLEM #1.6: A block of initial dimensions L_{xo}, L_{yo}, L_{zo} is deformed so that the new dimensions are L_x, L_y, L_z. Express the volume strain, $\ln(V/V_o)$, in terms of the three true strains, ε_x, ε_y, ε_z.

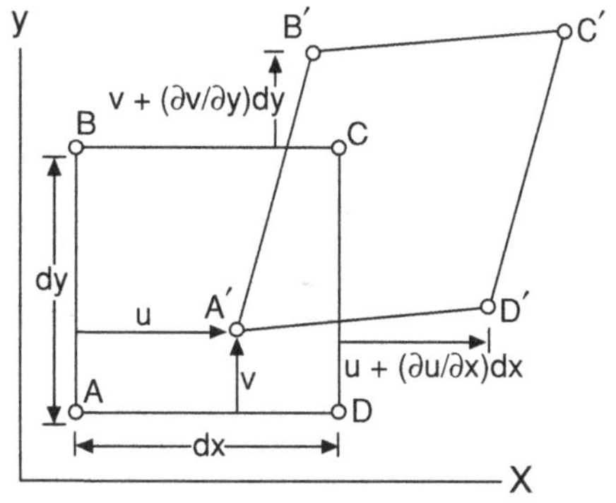

Figure 1.10. Translation, rotation, and distortion of a two-dimensional body.

Solution: (a) $V/V_o = L_x L_y L_z/(L_{xo} L_{yo} L_{zo})$, so $\ln(V/V_o) = \ln(L_x/L_{xo}) + \ln(L_y/L_{yo}) + \ln(L_z/L_{zo}) = \varepsilon_x + \varepsilon_y + \varepsilon_z$.

Note that if there is no volume change, $\ln(V/V_o) = 0$, and the sum of the normal strains is

$$\varepsilon_x + \varepsilon_y + \varepsilon_z = 0.$$

Small Strains

As bodies deform, they often undergo translations and rotations as well as deformation. Strain must be defined in such a way as to exclude the effects of translation and rotation. Consider a two-dimensional body in Figure 1.10. Normal strains are defined as the fractional extensions (tensile) or contractions (compressive), $\Delta L/L_o$, so $\varepsilon_{xx} = (\overline{A'D'} - \overline{AD})/\overline{AD} = \overline{A'D'}/\overline{AD} - 1$. For a small strain, this reduces to

$$\varepsilon_{xx} = (\partial u/\partial x)dx/dx = \partial u/\partial x. \tag{1.21}$$

Similarly,

$$\varepsilon_{yy} = (\partial v/\partial y)/dy/dy = \partial v/\partial y. \tag{1.22}$$

Shear strains are similarly defined in terms of the angles between AD and $A'D'$ and between AB and $A'B'$, which are respectively

$$(\partial v/\partial x)dx/dx = \partial v/\partial x \quad \text{and} \quad (\partial u/\partial y)dy/dy = \partial u/\partial y.$$

The total engineering shear strain, γ_{yx}, is the sum of these angles,

$$\gamma_{yx} = \partial v/\partial x + \partial u/\partial y = \gamma_{xy}. \tag{1.23}$$

Figure 1.11 shows that this definition excludes the effects of rotation for small strains.

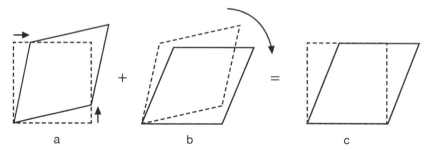

Figure 1.11. Illustration of shear and rotation. With small deformations, (a) differs from (c) only by a rotation, (b).

For a three-dimensional body with displacements w in the z-direction,

$$\varepsilon_{yy} = \partial v/\partial y, \quad \gamma_{yz} = \gamma_{zy} = \partial w/\partial y + \partial v/\partial z, \quad \text{and} \quad \gamma_{yx} = \gamma_{xy} = \partial v/\partial x + \partial u/\partial y$$

Small strains can be treated as tensors,

$$\varepsilon_{ij} = \begin{vmatrix} \varepsilon_{xx} & \varepsilon_{yx} & \varepsilon_{zx} \\ \varepsilon_{xy} & \varepsilon_{yy} & \varepsilon_{yz} \\ \varepsilon_{xz} & \varepsilon_{yz} & \varepsilon_{zz} \end{vmatrix}, \tag{1.24}$$

where the mathematical shear strains, ε_{ij}, are one half of the engineering shear strains, γ_{ij}:

$$\varepsilon_{yz} = \varepsilon_{zy} = (1/2)\gamma_{yz} = (1/2)(\partial v/\partial z + \partial w/dy)$$
$$\varepsilon_{zx} = \varepsilon_{xz} = (1/2)\gamma_{zx} = (1/2)(\partial w/\partial x + \partial u/dz \tag{1.25}$$
$$\varepsilon_{xy} = \varepsilon_{yx} = (1/2)\gamma_{xy} = (1/2)(\partial u/\partial y + \partial v/\partial x).$$

Transformation of Axes

Small strains may be transformed from one set of axes to another in a manner completely analogous to the transformation of stresses (see Equation 1.9),

$$\varepsilon_{ij} = \ell_{im}\ell_{jn}\varepsilon_{mn}, \tag{1.26}$$

where double summation is implied. For example,

$$\begin{aligned} \varepsilon_{x'x'} = \ &\ell_{x'x}\ell_{x'x}\varepsilon_{xx} + \ell_{x'y}\ell_{x'x}\varepsilon_{yx} + \ell_{x'z}\ell_{x'x}\varepsilon_{zx} \\ &+ \ell_{x'x}\ell_{x'y}\varepsilon_{xy} + \ell_{x'y}\ell_{x'y}\varepsilon_{yy} + \ell_{x'z}\ell_{x'y}\varepsilon_{zy} \\ &+ \ell_{x'x}\ell_{x'z}\varepsilon_{xz} + \ell_{x'y}\ell_{x'z}\varepsilon_{yz} + \ell_{x'z}\ell_{x'z}\varepsilon_{zz} \end{aligned} \tag{1.27a}$$

and

$$\begin{aligned} \varepsilon_{x'y'} = \ &\ell_{x'x'}\ell_{y'x}\varepsilon_{xx} + \ell_{x'y}\ell_{y'x}\varepsilon_{yx} + \ell_{x'z}\ell_{y'x}\varepsilon_{zx} \\ &+ \ell_{x'x}\ell_{y'y}\varepsilon_{xy} + \ell_{x'y}\ell_{y'y}\varepsilon_{yy} + \ell_{x'z}\ell_{y'y}\varepsilon_{zy} \\ &+ \ell_{x'x}\ell_{y'z}\varepsilon_{xz} + \ell_{x'y}\ell_{y'z}\varepsilon_{yz} + \ell_{x'z}\ell_{y'z}\varepsilon_{zz}. \end{aligned} \tag{1.27b}$$

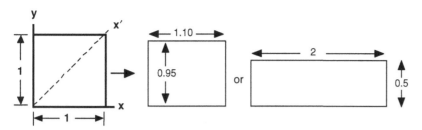

Figure 1.12. Small and large deformations of a body.

These can be written more simply in terms of the usual shear strains,

$$\begin{aligned}
\varepsilon_{x'} = &\, \ell_{x'x}^2 \varepsilon_x + \ell_{x'y}^2 \varepsilon_y + \ell_{x'z}^2 \varepsilon_z \\
&+ \ell_{x'y}\ell_{x'z}\gamma_{yz} + \ell_{x'z}\ell_{x'x}\gamma_{zx} + \ell_{x'x}\ell_{x'y}\gamma_{xy}
\end{aligned} \tag{1.28a}$$

and

$$\begin{aligned}
\gamma_{x'y'} = &\, 2(\ell_{x'x}\ell_{y'x}\varepsilon_x + \ell_{x'y}\ell_{y'y}\varepsilon_y + \ell_{x'z}\ell_{y'z}\varepsilon_z) \\
&+ (\ell_{x'y}\ell_{y'z} + \ell_{x'z}\ell_{y'y})\gamma_{yz} + (\ell_{x'z}\ell_{y'x} + \ell_{x'x}\ell_{y'z})\gamma_{zx} \\
&+ (\ell_{x'x}\ell_{y'y} + \ell_{x'y}\ell_{y'x})\gamma_{xy}.
\end{aligned} \tag{1.28b}$$

Large strains are not tensors and cannot be transformed from one axis system to another by a tensor transformation. The reason is that the angles between material directions are altered by deformation. With small strains, changes of angle are small and can be neglected. The following example illustrates this point.

EXAMPLE PROBLEM #1.7: A two-dimensional square body initially 1.000 cm by 1.000 cm was deformed into a rectangle 0.95 cm by 1.10 cm as shown in Figure 1.12.

a. Calculate the strain, $e_{x'}$, along the diagonal from its initial and final dimensions. Then calculate the strains, e_x and e_y, along the edges and use the transformation equation, $e_{ij} = \Sigma\Sigma \ell_{im}\ell_{jn}e_{mn}$, to find the strain along the diagonal and compare with the two values of $e_{x'}$.
b. Repeat (a) for a 1.000 cm by 1.000 cm square deformed into a 0.50 cm by 2.0 cm rectangle.

Solution: (a) The initial diagonal is $\sqrt{2} = 1.414214$, and for the small deformation the final diagonal becomes $\sqrt{(0.95^2 + 1.1^2)} = 1.4534$, so $e_{x'} = (L - L_o)/L_o = L/L_o - 1 = 1.4534/1.414214 - 1 = 0.0277$.

Taking the angle, θ, between the x' and x (or y) axes as $45°$, $e_{x'} = \ell_{x'x}^2 e_x + \ell_{x'y}^2 e_y = 0.1/2 - 0.05/2 = 0.025$, which is very close to 0.0277.
(b) For the large deformation, the diagonal becomes $\sqrt{(2^2 + 0.5^2)} = 2.062$ so calculating the strain from this, $e_{x'} = 2.062/1.414214 - 1 = 0.4577$.

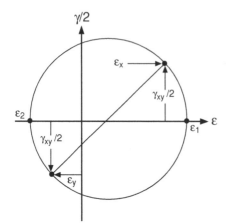

Figure 1.13. Mohr's circle for strain. This is similar to the Mohr's stress circle except that the normal strain, ε (or e), is plotted on the horizontal axis (instead of σ) and $\gamma/2$ is plotted vertically (instead of τ).

The strains on the edges are $e_x = 1$ and $e_y = -0.5$, so $e_{x'} = \ell_{x'x}^2 e_x + \ell_{x'y^2} e_y = (1/2)(1) + (1/2)(-.5) = 0.0250$, which does not agree with $e_{x'} = 0.4577$ calculated from the specimen dimensions. With true strains, the agreement is not much better. Direct calculation of $e_{x'}$ from the diagonal gives $e_{x'} = \ln(2.062/1.414) = 0.377$, and calculation from the strains along the sides gives $e_{x'} = (1/2)\ln(2) + (1/2)\ln(.5) = 0$. The reason is that with large strains, the angle θ changes with deformation.

Mohr's Strain Circles

Because small strains are tensor quantities, Mohr's circle diagrams apply if the strains e_x, e_y, and γ_{xy} are known along two axes and the third axis, z, is a principal strain axis ($\gamma_{yz} = \gamma_{zx} = 0$). The strains may be either engineering, e, or true, ε, because they are equal when small. Care must be taken to remember that the tensor shear-strain terms are only one half of the conventional shear strains. A plot of $\gamma/2$ vs. e (or ε) is a circle, as shown in Figure 1.13. The equations, analogous to those for stresses, are

$$(e_1 + e_2)/2 = (e_x + e_y)/2 \tag{1.29}$$

$$(e_1 - e_2)/2 = \{[(e_x - e_y)/2]^2 + (\gamma_{xy}/2)\}^{1/2} \tag{1.30}$$

$$e_1, e_2 = (e_x + e_y)/2 \pm \{[(e_x - e_y)/2]^2 + (\gamma_{xy}/2)^2\}^{1/2} \tag{1.31}$$

$$\tan(2\theta) = (\gamma_{xy}/2)/[(e_x - e_y)/2] = \gamma_{xy}/(e_x - e_y) \tag{1.32}$$

As with Mohr's stress circles, a three-dimensional strain state can be represented by three Mohr's circles. It is emphasized that the strain transformation equations, including the Mohr's circle equations, apply to small strains. Errors increase when the strains are large enough to cause rotation of the axes.

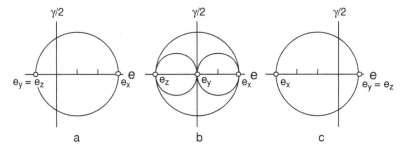

Figure 1.14. Three-dimensional Mohr's strain circles for (A) tension, (B) plane strain, (C) compression. In A and C, the circles between e_y and e_z reduce to a point and the circles between e_x and e_y coincide with the circles between e_x and e_z.

EXAMPLE PROBLEM #1.8: Draw the three-dimensional Mohr's circle diagram for an x-direction tension test (assume $e_y = e_z = -e_x/3$), plane strain ($e_y = 0$), and an x-direction compression test (assume $e_y = e_z = -e_x/3$). In each case, determine γ_{max}.

Solution: The three-dimensional Mohr's circle diagrams are shown in Figure 1.14.

Force and Moment Balances

The solutions of many mechanics problems require force and moment balances. The external forces acting on one half of the body must balance those acting across the cut. Consider force balances to find the stresses in the walls of a capped thin-wall tube loaded by internal pressure.

EXAMPLE PROBLEM #1.9: A capped thin-wall tube having a length, L, a diameter, D, and a wall thickness, t, is loaded by an internal pressure P. Find the stresses in the wall assuming that t is much smaller than D and that D is much less than L.

Solution: First make a cut perpendicular to the axis of the tube (Figure 1.15a), and consider the vertical (y-direction) forces. The force from the pressurization is the pressure, P, acting on the end area of the tube, $P\pi D^2/4$. The force in the wall is the stress, σ_y, acting on the cross-section of the wall, $\sigma_y\pi Dt$. Balancing the forces, $P\pi D^2/4 = \sigma_y\pi Dt$, and solving for σ_y,

$$\sigma_y = PD/(4t) \tag{1.33}$$

The hoop stress, σ_x, can be found from a force balance across a vertical cut (Figure 15b). The force acting to separate the tube is the pressure, P, acting on the internal area, DL, where L is the tube length. This is balanced

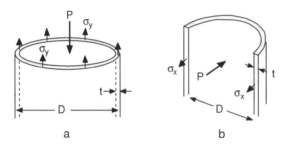

Figure 1.15. Cuts through a thin-wall tube loaded under internal pressure.

a   b

by the hoop stress in the two walls, σ_x, acting on the cross-sectional area of the two walls, $2Lt$. (The force in the capped ends is neglected because the tube is long and we are interested in the stress in regions remote from the ends.) Equating these two forces, $PDL = \sigma_x 2t$,

$$\sigma_x = PD/(2t) = 2\sigma_y. \tag{1.34}$$

The stress in the radial direction though the wall thickness, σ_z, is negligible relative to σ_x and σ_y if $D \gg t$. On the inside surface, $\sigma_z = -P$, and on the outside surface, $\sigma_z = 0$. The average, $-P/2$, is much less than $PD/(2t)$, so for engineering purposes we can take $\sigma_z = 0$.

A moment balance may be made about any axis in a body under equilibrium. The moment caused by external forces must equal the moment caused by internal stresses. An example is the torsion of a circular rod.

EXAMPLE PROBLEM #1.10: Relate the internal shear stress, τ_{xy}, in a rod of radius, R, to the torque, T, acting on the rod.

Solution: Consider a differential element of dimensions $(2\pi r)(dr)$ in a tubular element at a radius, r, from the axis and of thickness dr (Figure 1.16). The shear force acting on this element is the shear stress multiplied by the area, $\tau_{xy}(2\pi r)(dr)$. The torque about the central axis caused by this element is the shear force multiplied by the distance from the axis, r, $dT = \tau_{xy}(2\pi r)(r)dr$, so

$$T = 2\pi \int_0^R \tau_{xy} r^2 dr \tag{1.35}$$

An explicit solution requires knowledge of how τ_{xy} varies with r.

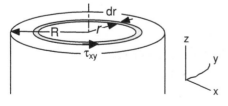

Figure 1.16. A differential element of area $2\pi r dr$ in a rod loaded under torsion. The shear stress, τ_{xy}, on this element causes a differential moment, $2\pi \tau_{xy} r^2 dr$.

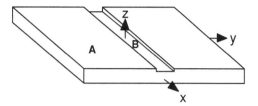

Figure 1.17. Grooved plate. The material outside the groove affects the x-direction flow inside the groove, so $\varepsilon_{xA} = \varepsilon_{xB}$.

Common Boundary Conditions

It is important in analyzing mechanical problems to recognize simple, often unstated, boundary conditions and use them to make simplifying assumptions.

1) Free surfaces: On a free surface, the two shear stress components in the surface vanish, that is, if z is the normal to a free surface, $\tau_{yz} = \tau_{zx} = 0$. Unless there is a pressure acting on a free surface, the stress normal to it also vanishes, that is, $\sigma_{zz} = 0$. Likewise, there are no shear stresses, $\tau_{yz} = \tau_{zx}$, acting on surfaces that are assumed to be frictionless.

2) Constraints from neighboring regions: The deformation in a particular region is often controlled by the deformation in a neighboring region. Consider the deformation in a long narrow groove in a plate as shown in Figure 1.17. The long narrow groove (B) is in close contact with a thicker region (A). As the plate deforms, the deformation in the groove must be compatible with the deformation outside the groove. Its elongation or contraction must be the same as that in the material outside so that $\varepsilon_{xA} = \varepsilon_{xB}$. The y- and z-direction strains in the two regions need not be the same.

3) St. Venant's principle states that the restraint from any end or edge effect will disappear within one characteristic distance. As an example, the enlarged end of a tensile bar (Figure 1.18) tends to suppress lateral contraction of the gauge section next to it. Here the characteristic distance is the diameter of the gauge section, so the constraint is almost gone at a distance from the enlarged end equal to the diameter.

Another example of St. Venant's principle is a thin, wide sheet bent to a constant radius of curvature (Figure 1.20). The condition of plane strain ($\varepsilon_y = 0$) prevails over most of the material. This is because the top and bottom surfaces are so close that they restrain one another from contracting or expanding. Appreciable deviation from plane strain occurs only within a distance from the edges of the sheet equal to the sheet thickness. At the edge, $\sigma_y = 0$.

Figure 1.18. In the gauge section of a tensile bar, the effect of the ends almost disappears at a distance, d, from the shoulder.

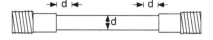

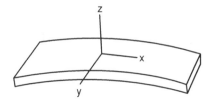

Figure 1.19. Bending of a thin sheet. Plane strain ($\varepsilon_y = 0$) prevails except near the edges, where there is a condition of plane stress ($\sigma_V = 0$).

Note

The simplification of the notation for tensor transformation of $\sigma_{ij} = \sum_{n=1}^{3} \sum_{m=1}^{3} \ell_{im}\ell_{jn}\sigma_{mn}$ to $\sigma_{ij} = \ell_{im}\ell_{jn}\sigma_{mn}$ has been attributed to Albert Einstein.

Problems

1. A body is loaded under a stress state, $\sigma_x = 400$, $\sigma_y = 100$, $\tau_{xy} = 120$, $\tau_{yz} = \tau_{zx} = \sigma_z = 0$.
   a. Sketch the Mohr's circle diagram.
   b. Calculate the principal stresses.
   c. What is the largest shear stress in the body? (Do not neglect the z direction.)

2. Three strain gauges have been pasted on the surface of a piece of steel in the pattern shown in Figure 1.20. While the steel is under load, these gauges indicate the strains parallel to their axes:

   Gauge $A = 450 \times 10^{-6}$: Gauge $B = 300 \times 10^{-6}$: Gauge $C = 150 \times 10^{-6}$
   a. Calculate the principal strains ε_1 and ε_2.
   b. Find the angle between the 1-axis and the x-axis, where 1 is the axis of the largest principal strain. [Hint: Let the direction of Gauge B be x', write the strain transformation equation expressing the strain $e_{x'}$ in terms of the strains along the x-y axes, solve for γ_{xy}, and finally use the Mohr's circle equations.]

3. Consider a thin-wall tube that is 1 in. in diameter and has a 0.010 in. wall thickness. Let x, y, and z be the axial, tangential (hoop), and radial directions, respectively.

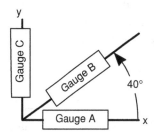

Figure 1.20. Arrangement of strain gauges.

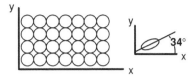

Figure 1.21. Circle grids printed on a metal sheet.

a. The tube is subjected to an axial tensile force of 80 lbs and a torque of 100 in.-lbs.
   i. Sketch the Mohr's stress circle diagram showing stresses in the x-y plane.
   ii. What is the magnitude of the largest principal stress?
   iii. At what angles are the principal stress axes, 1 and 2, to the x and y directions?
b. Now let the tube be capped and let it be subject to an internal pressure of 120 psi and a torque of 100 in.-lbs.
   i. Sketch the Mohr's stress circle diagram showing stresses in the x-y plane.
   ii. What is the magnitude of the largest principal stress?
   iii. At what angles are the principal stress axes, 1 and 2, to the x and y directions?

4. A solid is deformed under plane-strain conditions ($\varepsilon_z = 0$). The strains in the x-y plane are $\varepsilon_x = 0.010$, $\varepsilon_y = 0.005$, and $\gamma_{xy} = 0.007$.
   a. Sketch the Mohr's strain circle diagram.
   b. Find the magnitude of ε_1 and ε_2.
   c. What is the angle between the 1- and x-axes?
   d. What is the largest shear strain in the body? (Do not neglect the z direction.)

5. A grid of circles, each 10.00 mm in diameter, was etched on the surface of a sheet of steel (Figure 1.21). When the sheet was deformed, the grid circles were distorted into ellipses. Measurement of one indicated that the major and minor diameters were 11.54 mm and 10.70 mm, respectively.
   a. What are the principal strains, ε_1 and ε_2?
   b. If the axis of the major diameter of the ellipse makes an angle of 34° with the x-direction, what is the shear strain, γ_{xy}?
   c. Draw the Mohr's strain circle showing ε_1, ε_2, ε_x, and ε_y.

7. Consider the torsion of a rod that is 1 m long and 50 mm in diameter.
   a. If one end of the rod is twisted by 1.2° relative to the other end, what would be the largest principal strain on the surface?

Figure 1.22. Glued rod.

b. If the rod were extended by 1.2% and its diameter decreased by 0.4% at the same time it was being twisted, what would be the largest principal strain?

8. Two pieces of rod are glued together along a joint whose normal makes an angle, θ, with the rod axis, x, as shown in Figure 1.22. The joint will fail if the shear stress on the joint exceeds its shear strength, τ_{max}. It will also fail if the normal stress across the joint exceeds its normal strength, σ_{max}. The shear strength, τ_{max}, is 80% of the normal strength, σ_{max}. The rod will be loaded in uniaxial tension along its axis, and it is desired that the rod carry as high a tensile force, F_x, as possible. The angle, θ, cannot exceed 65°.

a. At what angle, θ, should the joint be made so that a maximum force can be carried?

b. If θ_{max} were limited to 45° instead of 65°, how would your answer be altered? [Hint: Plot σ_x/σ_{max} vs. θ for both failure modes.]

9. Consider a tube made by coiling and gluing a strip as show in Figure 1.23. The diameter is 1.5 in., the length is 6 in., and the wall thickness is 0.030 in. If a tensile force of 80 lbs and a torque of 30 in.-lbs are applied in the direction shown, what is the stress normal to the glued joint? [Hint: Set up a coordinate system.]

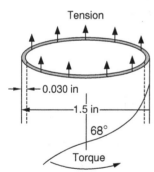

Figure 1.23. Tube formed from a coiled and glued strip.

2 Elasticity

Introduction

Elastic deformation is reversible. When a body deforms elastically under a load, it will revert to its original shape as soon as the load is removed. A rubber band is a familiar example. However, most materials can undergo very much less elastic deformation than rubber. In crystalline materials elastic strains are small, usually less than 0.5%. For most materials other than rubber, it is safe to assume that the amount of deformation is proportional to the stress. This assumption is the basis of the following treatment. Because elastic strains are small, it does not matter whether the relations are expressed in terms of engineering strains, e, or true strains, ε. The treatment in this chapter covers elastic behavior of isotropic materials, the temperature dependence of elasticity, and thermal expansion. Anisotropic elastic behavior is covered in Chapter 15.

Isotropic Elasticity

An *isotropic* material is one that has the same properties in all directions. If uniaxial tension is applied in the x-direction, the tensile strain is $\varepsilon_x = \sigma_x/E$, where E is *Young's modulus*. Uniaxial tension also causes lateral strains $\varepsilon_y = \varepsilon_z = -\upsilon\varepsilon_x$, where υ is *Poisson's ratio*. Consider the strain, ε_x, produced by a general stress state, $\sigma_x, \sigma_y, \sigma_z$. The stress, σ_x, causes a contribution $\varepsilon_x = \sigma_x/E$. The stresses σ_y, σ_z cause Poisson contractions $\varepsilon_x = -\upsilon\sigma_y/E$ and $\varepsilon_x = -\upsilon\sigma_z/E$. Taking into account these Poisson contractions, the general statement of *Hooke's law* is

$$\varepsilon_x = (1/E)[\sigma_x - \upsilon(\sigma_y + \sigma_z)]. \tag{2.1a}$$

Shear strains are affected only by the corresponding shear stress, so

$$\gamma_{yz} = \tau_{yz}/G = 2\varepsilon_{yz}, \tag{2.1b}$$

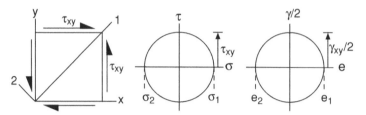

Figure 2.1. Mohr's stress and strain circles for shear.

where G is the shear modulus. Similar expressions apply for all directions:

$$\varepsilon_x = (1/E)[\sigma_x - \upsilon(\sigma_y + \sigma_z)] \qquad\qquad \gamma_{yz} = \tau_{yz}/G$$
$$\varepsilon_y = (1/E)[\sigma_y - \upsilon(\sigma_z + \sigma_x)] \qquad\qquad \gamma_{zx} = \tau_{zx}/G \qquad (2.2)$$
$$\varepsilon_z = (1/E)[\sigma_z - \upsilon(\sigma_x + \sigma_y)] \qquad\qquad \gamma_{xy} = \tau_{xy}/G.$$

Equations 2.1 and 2.2 hold whether or not the x, y, and z directions are directions of principal stress.

For an isotropic material, the shear modulus, G, is not independent of E and υ. This can be demonstrated by considering a state of pure shear, τ_{xy}, with $\sigma_x = \sigma_y = \sigma_z = \tau_{yz} = \tau_{zx} = 0$. The Mohr's circle diagram, Figure 2.1, shows that the principal stresses are $\sigma_1 = \tau_{xy}$, $\sigma_2 = -\tau_{xy}$, and $\sigma_3 = 0$.

From Hooke's law, $\varepsilon_1 = (1/E)[\sigma_1 - \upsilon(\sigma_2 + \sigma_3)] = (1/E)[\tau_{xy} - \upsilon(-\tau_{xy} + 0)] = [(1 + \upsilon)/E]\tau$. The Mohr's strain circle diagram (Figure 2.1) shows that $\gamma_{xy}/2 = \varepsilon_1$. Substituting for σ_1, $\gamma_{xy}/2 = [(1 + \upsilon)/E]\tau_{xy}$. Now comparing with $\gamma_{xy} = \tau_{xy}/G$,

$$G = E/[2(1 + \upsilon)]. \qquad (2.3)$$

The bulk modulus, B, is defined by the relation between the volume strain and the mean stress,

$$\Delta V/V = (1/B)\sigma_m, \qquad (2.4)$$

where $\sigma_m = (\sigma_x + \sigma_y + \sigma_z)/3$. Like the shear modulus, the bulk modulus is not independent of E and υ. This can be demonstrated by considering the volume strain produced by a state of hydrostatic stress, $\sigma_x = \sigma_y = \sigma_z = \sigma_m$. It was shown in Example Problem 1.7 that for small deformations, the volume strain $\Delta V/V = \varepsilon_x + \varepsilon_y + \varepsilon_z$. Substituting $\varepsilon_x = (1/E)[\sigma_x - \upsilon(\sigma_y + \sigma_z)] = (1/E)[\sigma_m - \upsilon(\sigma_m + \sigma_m)] = [(1 - 2\upsilon)/E]\sigma_m$. Then $\Delta V/V = 3\sigma_m(1 - 2\upsilon)/E$. Comparing with $\Delta V/V = (1/B)\sigma_m$,

$$B = E/[3(1 - 2\upsilon)]. \qquad (2.5)$$

For most materials, Poisson's ratio is between 0.2 and 0.4. The value of υ for an isotropic material cannot exceed 0.5; if υ were larger than 0.5, B would be negative, which would imply an increase of pressure would cause an increase of volume.

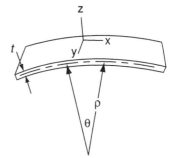

Figure 2.2. Bent sheet. The strain e_y and the stress σ_z are zero.

Any of the four elastic constants, E, υ, G, or B can be expressed in terms of two others:

$$E = 2G(1 + \upsilon) = 3B(1 - 2\upsilon) = 9BG/(G + 3B) \tag{2.6a}$$

$$\upsilon = E/2G - 1 = 1/2 - E/6B) = 1/2 - (3/2)G/(3B + G) \tag{2.6b}$$

$$G = E/[2(1 + \upsilon)] = 3EB/(9B - E) = (3/2)E(1 - 2\upsilon)/(1 + \upsilon) \tag{2.6c}$$

$$B = E/[3(1 - 2\upsilon)] = EG/[3(3G - E)] = (2/3)G(1 + \upsilon)/(1 - 2\upsilon). \tag{2.6d}$$

Only three of the six normal stresses and strain components, e_x, e_y, e_z, σ_x, σ_y, or σ_z, need be known to find the other three. This is illustrated in the following examples.

EXAMPLE PROBLEM #2.1: A wide sheet (1 mm thick) of steel is bent elastically to a constant radius of curvature, $\rho = 50$ cm, measured from the axis of bending to the center of the sheet, as shown in Figure 2.2. Knowing that $E = 208$ GPa and $\upsilon = 0.29$ for steel, find the stress in the surface. Assume that there is no net force in the plane of the sheet.

Solution: Designate the directions as shown in Figure 2.2. Inspection shows that $\sigma_z = 0$ because the stress normal to a free surface is zero. Also $e_y = 0$ because the sheet is wide relative to its thickness and e_y must be the same on top and bottom surfaces, therefore it equals zero. Finally, the strain e_x at the outer surface can be found geometrically as $e_x = (t/2)/r = (1/2)/500 = 0.001$. Now substituting into Hooke's laws (Equation 2.2), $e_y = (1/E)[\sigma_y - v(\sigma_x - 0)]$, so $\sigma_y = v\sigma_x$. Substituting again, $e_x = t/(2r) = (1/E)[\sigma_x - v(\sigma_y - 0)] = (1/E)(\sigma_x - v^2\sigma_x) = \sigma_x(1 - v^2)/E$. $\sigma_x = [t/(2r)]E/(1 - v^2)] = (0.001)(208 \times 10^9)/(1 - 0.29^2) = 227$ MPa. $\sigma_y = v\sigma_x = 0.29 \cdot 227 = 65.8$ MPa.

EXAMPLE PROBLEM #2.2: A body is loaded elastically so that $e_y = 0$ and $\sigma_z = 0$. Both σ_x and σ_y are finite. Derive an expression for e_z in terms of σ_x, E, and υ only.

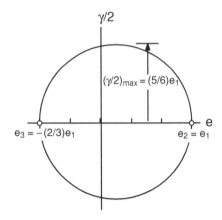

Figure 2.3. Mohr's strain circles for an elastic body under biaxial tension. Because the $(\sigma_1 - \sigma_2)$ circle reduces to a point, the $(\sigma_1 - \sigma_3)$ and $(\sigma_2 - \sigma_3)$ circles are superimposed as a single circle.

Solution: Substituting $e_y = 0$ and $\sigma_z = 0$ into Hooke's law (Equation 2.2),

$$0 = (1/E)[\sigma_y - \upsilon(0 + \sigma_x)], \quad \text{so } \sigma_y = \upsilon\sigma_x.$$

Substituting into Hooke's law again,

$$e_z = (1/E)[0 - \upsilon(\sigma_x + \sigma_x)] = -\sigma_x\upsilon(1 + \upsilon)/E.$$

EXAMPLE PROBLEM #2.3: Draw the three-dimensional Mohr's strain circle diagram for a body loaded elastically under balanced biaxial tension, $\sigma_1 = \sigma_2$. Assume $\upsilon = 0.25$. What is the largest shear strain, γ?

Solution: With $\sigma_1 = \sigma_2$ and $\sigma_3 = 0$, from symmetry, $e_1 = e_2$, so using Hooke's laws, $e_1 = (1/E)(\sigma_1 - \upsilon\sigma_1) = 0.75\sigma_1/E$, $e_3 = (1/E)[0 - \upsilon(\sigma_1 + \sigma_1)] = -0.5\sigma_1$. The maximum shear strain, $\gamma_{\max} = (e_1 - e_3) = 1.25\sigma_1/E$. In Figure 2.3, e_1, e_2, and e_3 are plotted on the horizontal axis. The circle has a radius of $0.625\sigma_1/E = (5/6)e_1$ with $e_2 = e_1$ and $e_3 = (-2/3)e_1$.

Variation of Young's Modulus

The binding energy between two neighboring atoms or ions can be represented by a potential well. Figure 2.4 is a schematic plot showing how the binding energy varies with the separation of atoms. At absolute zero, the equilibrium separation corresponds to the lowest energy. The Young's modulus and the melting point are both related to the potential well; the modulus depends on the curvature at the bottom of the well, and the melting point depends on the depth of the well. The curvature and the depth tend to be related, so the elastic moduli of different elements roughly correlate with their melting points (Figure 2.5). The modulus also correlates with the heat of fusion and the latent heat of melting, which are related to the depth of the potential well. For a given metal, the elastic modulus decreases as the temperature is increased

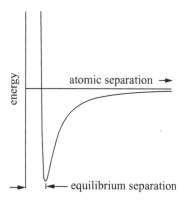

Figure 2.4. Schematic plot of the variation of binding energy with atomic separation at absolute zero.

from absolute zero to the melting point. This is illustrated for several metals in Figure 2.6. The value of the elastic modulus at the melting point is in the range of one-third and one-fifth its value at absolute zero.

For crystalline materials, the elastic modulus is generally regarded as being relatively structure-insensitive to changes of microstructure. Young's moduli of body-centered cubic (bcc) and face-centered cubic (fcc) forms of iron differ by a relatively small amount. Heat treatments that have large effects on hardness and yield strength, have little effect on the elastic properties. Cold working and small alloy additions also have relatively small effects because only a small fraction of the near-neighbor bonds are affected by these changes.

The elastic behavior of polymers is very different from that of metals. The elastic strains are largely a result of straightening polymer chains by rotation of

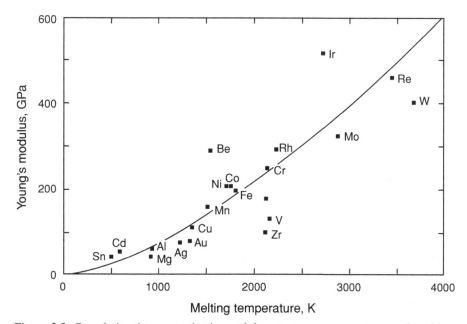

Figure 2.5. Correlation between elastic modulus at room temperature and melting point. Elements with higher melting points have greater elastic moduli.

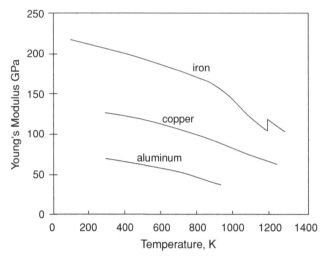

Figure 2.6. Variation of Young's modulus with temperature. Young's moduli decrease with increasing temperature. However, the moduli at the melting point are less than five times that at 0 K. Data from W. Koster, *Zeitschrift fur Metallkunde*, v. 39 (1948).

bonds rather than by bond stretching. Consequently, Young's moduli of typical polymers of are orders of magnitudes lower that those of metals and ceramics. Only in highly oriented polymers is the stretching of covalent bonds the primary source of elastic strains and the elastic moduli of highly oriented polymers are comparable with metals and ceramics. For polymers, the magnitude of elastic strain is often much greater than in metals and ceramics. Hooke's law does not describe the elasticity of rubber very well. This is treated in Chapter 12.

Isotropic Thermal Expansion

When the temperature of an isotropic material is changed, its fractional change in length is

$$\Delta L / L = \alpha \Delta T. \tag{2.7}$$

Such dimension changes can be considered strains, and Hooke's law can be generalized to include thermal strains:

$$e_x = (1/E)[\sigma_x - \upsilon(\sigma_y + \sigma_z)] + \alpha \Delta L. \tag{2.8}$$

This generalization is useful for finding the stresses that arise when constrained bodies are heated or cooled.

   EXAMPLE PROBLEM #2.4: A glaze is applied on a ceramic body by heating above it 600°C, which allows it to flow over the surface. On cooling,

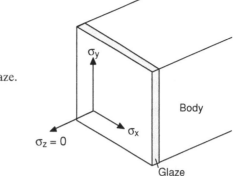

Figure 2.7. Stresses in a glaze.

the glaze becomes rigid at 500°C (see Figure 2.7). The coefficients of thermal expansion of the glaze and the body are $\alpha_g = 4.0 \times 10^{-6}/°C$ and $\alpha_b = 5.5 \times 10^{-6}/°C$. The elastic constants for the glaze are $E_g = 70\,\text{GPa}$ and $\upsilon_g = 0.3$. Calculate the stresses in the glaze when it has cooled to 20°C.

Solution: Let the direction normal to the surface be z so the x and y directions lie in the surface. The strains in the x and y directions must be the same in the glaze and the body,

$$e_{xg} = (1/E_g)[\sigma_{xg} - \upsilon_g(\sigma_{yg} + \sigma_{zg})] + \alpha_g \Delta T$$
$$= (1/E_b)[\sigma_{xb} - \upsilon_b(\sigma_{yb} + \sigma_{zb})] + \alpha_b \Delta T.$$

Taking $\sigma_{zg} = \sigma_{zb} = 0$ (free surface) and $\sigma_{xg} = \sigma_{yg}$ (symmetry), $\sigma_{xb} = \sigma_{yb} = 0$. Because the glaze is very thin, the stress in the ceramic body, $\sigma_{xb} = \sigma_{yb}$, required for force a balance, $t_b\sigma_{xb} + t_g\sigma_{xg} = 0$, is negligible. Hence,

$$(1/E_g)[\sigma_{xg}(1 - \upsilon_g)] + \alpha_g \Delta T = \alpha_b \Delta T,$$
$$\sigma_{xg} = [E_g/(1 - \upsilon_g)](\alpha_b - \alpha_g)\Delta T$$
$$= [(70 \times 10^9\text{Pa})/(1 - 0.3)]](4 - 5.5)(10^{-6}/°C)(-480°C)$$
$$= -72\text{MPa (compression) on cooling}$$

The elastic constants and thermal expansion coefficients for various materials are given in Table 2.1.

Notes

Robert Hooke (1635–1703) was appointed curator of experiments of the Royal Society on the recommendation of Robert Boyle. In that post, he devised apparati to clarify points of discussion as well as to demonstrate his ideas. He

Table 2.1. *Elastic constants and thermal expansion coefficient for various materials*

Material	E (GPa)	Poisson's ratio	Thermal expansion coefficient ($\times 10^{-6}/^\circ$C)
aluminum	62	0.24	23.6
iron	208.2	0.291	11.8
copper	128	0.35	16.5
magnesium	44	–	27.1
nickel	207	0.31	13.3
MgO	205	–	9
Al$_3$O$_3$	350	–	9
glass	70	–	9
polystyrene	2.8	–	63
PVC	0.35	–	190
nylon	2.8	–	100

noted that when he put weights on a string or a spring, its deflection was proportional to the weight. Hooke was a contemporary of Isaac Newton, whom he envied. Several times he claimed to have discovered things attributed to Newton but was never able to substantiate these claims.

Thomas Young (1773–1829) studied medicine but because of his interest in science, he took a post as lecturer at the Royal Institution. In his lectures, he was the first to define the concept of a modulus of elasticity, although he defined it quite differently than we do today. He described the modulus of elasticity in terms of the deformation at the base of a column due to its own weight.

Louis Navier (1785–1836) wrote a book on the strength of materials in 1826, in which he defined the elastic modulus as the load per unit cross-sectional area and determined *E* for iron.

Because his family was poor, S. D. Poisson (1781–1840) did not have a chance to learn to read or write until he was 15. After two years of visiting mathematics classes, he passed the exams for admission to École Polytechnique with distinction. He realized that axial elongation, *e*, must be accompanied by lateral contraction, which he took as $(-1/4)e$, not realizing that the ratio of contractile strain to elongation strain is a property that varies from material to material.

REFERENCES

H. B. Huntington, *The Elastic Constants of Crystals*, Academic Press (1964).
Simons & Wang, *Single Crystal Elastic Constants and Calculated Properties – A Handbook*, MIT Press (1971).
W. Köster and H. Franz, *Metals Review*, v. 6 (1961) pp. 1–55.

Problems

1. Reconsider Problem 2 in Chapter 1 and assume that $E = 205$ GPa and $\upsilon = 0.29$.
   A. Calculate the principal stresses under load.
   B. Calculate the strain, ε_z.
2. Consider a thin-walled tube, capped at each end and loaded under internal pressure. Calculate the ratio of the axial strain to the hoop strain, assuming that the deformation is elastic. Assume $E = 10^7$ psi and $\upsilon = 1/3$. Does the length of the tube increase, decrease, or remain constant?
3. A sheet of metal was deformed elastically under balanced biaxial tension $(\sigma_x = \sigma_y, \sigma_z = 0)$.
   A. Derive an expression for the ratio of elastic strains, e_z/e_x, in terms of the elastic constants.
   B. If E $= 70$ GPa and $\upsilon = 0.30$, and e_y is measured as 1.00×10^{-3}, what is the value of e_z?
4. A cylindrical plug of a gumite* is placed in a cylindrical hole in a rigid block of stiffite.* Then the plug is compressed axially (parallel to the axis of the hole). Assume that plug exactly fits the hole and that the stiffite does not deform at all. Assume elastic deformation and that Hooke's law holds. Derive an expression for the ratio of the axial strain to the axial stress, ε_a/σ_a, in the gummite in terms of the Young's modulus, E, and Poisson's ratio, υ, of the gummite.
5. Strain gauges mounted on a free surface of a piece of steel ($E = 205$ GPa, $\upsilon = 0.29$) indicate strains of $e_x = -0.00042$, $e_y = 0.00071$, and $\gamma_{xy} = 0.00037$.
   A. Calculate the principal strains.
   B. Use Hooke's laws to find the principal stresses from the principal strains.
   C. Calculate σ_x, σ_y, and τ_{xy} directly from e_x, e_y, and γ_{xy}.
   D. Calculate the principal stresses directly from σ_x, σ_y, and τ_{xy}, and compare your answers with the answers to B.
6. A steel block ($E = 30 \times 10^6$ psi and $\upsilon = 0.29$) is loaded under uniaxial compression along x.
   A. Draw the Mohr's strain circle diagram.
   B. There is an axis, x', along which the strain $\varepsilon_{x'} = 0$. Find the angle between x and x'.
7. Poisson's ratio for rubber is 0.5. What does this imply about the bulk modulus?

---

  * "Gummite" and "stiffite" are fictitious names.

Table 2.2. *Properties of Glass and PMMA*

Glass	PMMA	
thermal expansion coefficient (K^{-1})	$9. \times 10^{-6}$	$90. \times 10^{-6}$
Young's modulus (GPa)	69.	3.45
Poisson's ratio	0.28	.38

8. A plate of glass is sandwiched by two plates of polymethylmethacrylate, as shown in Figure 2.8. Assume that the composite is free of stresses at 40°C. Find the stresses when the assembly is cooled to 20°C. The properties of the glass and the polymethylmethacrylate are given here. The total thicknesses of the glass and the PMMA are equal. Assume each to be isotropic, and assume that creep is negligible. Properties of glass and PMMA are given in Table 2.2.

9. A bronze sleeve, 0.040 in thick, was mounted on a 2.000-in diameter steel shaft by heating it to 100°C while the temperature of the shaft was maintained at 20°C. Under these conditions, the sleeve just fit on the shaft with zero clearance. Find the principal stresses in the sleeve after it cools to 20°C. Assume that friction between the shaft and the sleeve prevented any sliding at the interface during cooling, and assume that the shaft is so massive and stiff that strains in the shaft itself are negligible. For bronze, $E = 16 \times 10^6$ psi, $v = 0.30$, and $\alpha = 18.4 \times 10^{-6} (°C)^{-1}$.

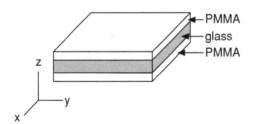

Figure 2.8. Laminated sheets of PMMA and glass.

3 Mechanical Testing

Introduction

Tensile properties are used in the selection of materials for various applications. Material specifications often include minimum tensile properties to assure quality, so tests must be made to insure that materials meet these specifications. Tensile properties are also used in research and development to compare new materials or processes. With plasticity theory (Chapter 5), tensile data can be used to predict a material's behavior under forms of loading other than uniaxial tension.

Often the primary concern is strength. The level of stress that causes appreciable plastic deformation is called its *yield stress*. The maximum tensile stress that a material carries is called its *tensile strength* (or *ultimate strength* or *ultimate tensile strength*). Both of these measures are used, with appropriate caution, in engineering design. A material's ductility may also be of interest. *Ductility* describes how much the material can deform before it fractures. Rarely, if ever, is the ductility incorporated directly into design. Rather, it is included in specifications only to assure quality and toughness. Elastic properties may be of interest but these usually are measured ultrasonically.

Tensile Testing

Figure 3.1 shows a typical tensile specimen. It has enlarged ends or shoulders for gripping. The important part of the specimen is the gauge section. The cross-sectional area of the gauge section is less than that of the shoulders and grip region so the deformation will occur here. The gauge section should be long compared to the diameter (typically four times). The transition between the gauge section and the shoulders should be gradual to prevent the larger ends from constraining deformation in the gauge section.

There are several ways of gripping specimens, as shown in Figure 3.2. The ends may be screwed into threaded grips, pinned, or held between wedges.

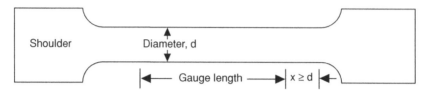

Figure 3.1. Typical tensile specimen with a reduced gauge section and larger shoulders.

Special grips are used for specimens with butt ends. The gripping system should assure that the slippage and deformation in the grip region is minimized. They should also prevent bending.

Figure 3.3 is a typical engineering stress-strain curve for a ductile material. For small strains the deformation is elastic and reverses if the load is removed. At higher stresses, plastic deformation occurs. This is not recovered when the load is removed. Bending a wire or paper clip with the fingers (Figure 3.4) illustrates the difference. If the wire is bent a small amount, it will snap back when released. However, if the bend is more severe it will only partly

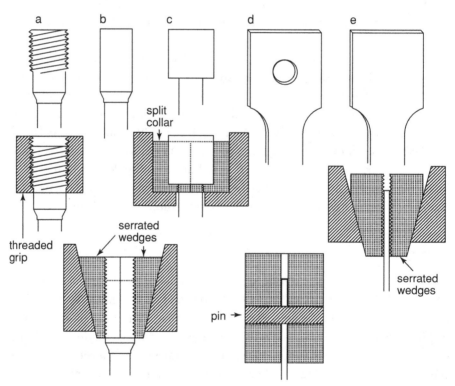

Figure 3.2. Systems for gripping tensile specimens. For round specimens these include (a) threaded grips, (b) serrated wedges, and (c) split collars for butt-end specimens. Sheet specimens may be gripped by (d) pins or by (e) serrated wedges. From W. F. Hosford in *Tensile Testing*, ASM Int. (1992).

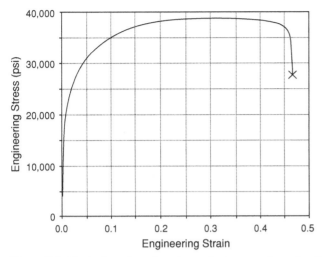

Figure 3.3. Typical engineering stress-strain curve for a ductile material.

recover, leaving a permanent bend. The onset of plastic deformation is usually associated with the first deviation of the stress-strain curve from linearity.*

It is tempting to define an *elastic limit* as the stress that causes the first plastic deformation and to define a *proportional limit* as the first departure from linearity. However, neither definition is very useful because they both depend on how accurately strain is measured. The more accurate the strain measurement is, the lower is the stress at which plastic deformation and nonlinearity can be detected.

To avoid this problem, the onset of plasticity is usually described by an *offset yield strength*. It is found by constructing a straight line parallel to the initial linear portion of the stress-strain curve but offset from it by $e = 0.002$ (0.2%.) The offset yield strength is taken as the stress level at which this straight line

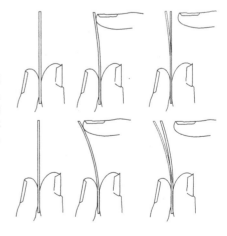

Figure 3.4. Using the fingers to sense the elastic and plastic responses of a wire. With a small force (top), all bending is elastic and disappears when the force is released. With a greater force (below), the elastic part of the bending is recoverable but the plastic part is not. From Hosford, *ibid*.

---

* For some materials, there may be nonlinear elastic deformation.

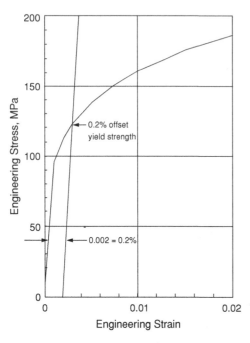

Figure 3.5. The low strain region of the stress-strain curve for a ductile material. From Hosford, *ibid*.

intersects the stress-strain curve (Figure 3.5). The rationale is that if the material had been loaded to this stress and then unloaded, the unloading path would have been along this offset line, resulting in a plastic strain of $e = 0.002$ (0.2%). The advantage of this way of defining yielding is that it is easily reproduced. Occasionally, other offset strains are used and or yielding is defined in terms of the stress necessary to achieve a specified total strain (e.g., 0.5%) instead of a specified plastic strain. In all cases, the criterion should be made clear to the user of the data.

Yield points: The stress–strain curves of some materials (e.g., low carbon steels and linear polymers) have an initial maximum followed by lower stress, as shown in Figures 3.6a and 3.6b. At any given instant after the initial maximum, all of the deformation occurs within a relatively small region of the specimen. For steels, this deforming region is called a *Lüder's band*. Continued elongation occurs by movement of the Lüder's band along the gauge section rather than by continued deformation within the band. Only after the band has traversed the entire gauge section does the stress rise again.

In the case of linear polymers, the yield strength is usually defined as the initial maximum stress. For steels, the subsequent lower yield strength is used to describe yielding. Because the initial maximum stress is extremely sensitive to the alignment of the specimen, it is not useful in describing yielding. Even so, the lower yield strength is sensitive to the strain rate. ASTM standards should be followed. The stress level during Lüder's band propagation fluctuates. Some laboratories report the minimum level as the yield strength and other use the average level.

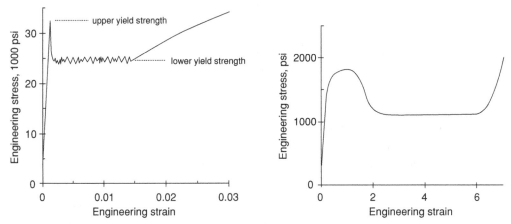

Figure 3.6. Inhomogeneous yielding of low-carbon steel (left) and a linear polymer (right). After the initial stress maximum, the deformation in both materials occurs within a narrow band that propagates the length of the gauge section before the stress rises again.

As long as the engineering stress-strain curve rises, the deformation will occur uniformly along the gauge length. For a ductile material, the stress will reach a maximum well before fracture. When the maximum is reached, the deformation localizes forming a neck as shown in Figure 3.7.

The *tensile strength* (or ultimate strength) is defined as the highest value of the engineering stress (Figure 3.8). For ductile materials, the tensile strength corresponds to the point at which necking starts. Less ductile materials fracture before they neck. In this case, the fracture stress is the tensile strength. Very brittle materials (e.g., glass) fracture before they yield. Such materials have tensile strengths but no yield stresses.

Ductility

Two common parameters are used to describe the ductility of a material. One is the *percent elongation,* which is simply defined as

$$\% \, El = (L_f - L_o)/L_o \times 100\%, \tag{3.1}$$

Figure 3.7. After a maximum of the stress strain curve, deformation localizes to form a neck.

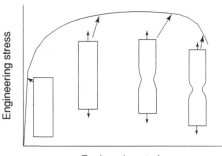

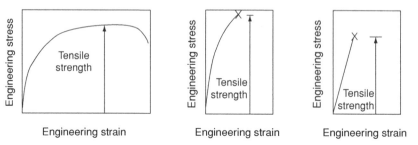

Figure 3.8. The tensile strength is the maximum engineering stress, regardless of whether the specimen necks or it fractures before necking. From Hosford, *ibid.*

where L_o is the initial gauge length and L_f is the length of the gauge section at fracture. Measurements may be made on the broken pieces or under load. For most materials, the elastic elongation is so small compared to the plastic elongation that it can be neglected. When this is not so, as with brittle materials or with rubber-like materials, it should be made clear whether or not the percent elongation includes the elastic portion.

The other common measure of ductility is the *percent reduction of area* at fracture, defined as

$$\% RA = (A_o - A_f)/A_o \times 100\%, \tag{3.2}$$

where A_o is the initial cross-sectional area and A_f is the cross-sectional area of the fracture. If the failure occurs before the necking, the $\% El$ can be calculated from the $\% RA$ by assuming constant volume. In this case,

$$\% El = \% RA/(100 - \% RA) \times 100\% \tag{3.3}$$

The $\% El$ and $\% RA$ are no longer directly related after a neck has formed.

As a measure of ductility, percent elongation has the disadvantage that it combines the uniform elongation that occurs before necking and the localized elongation that occurs during necking. The uniform elongation depends how the material strain hardens rather than the fracture behavior. The necking elongation is sensitive to the specimen shape. With a gauge section that is very long compared to the diameter, the contribution of necking to the total elongation is very small. On the other hand, if the gauge section is very short the necking elongation accounts for most of the elongation. For round bars, this problem has been remedied by standardizing the ratio of the gauge length to diameter at 4:1. However, there is no simple relation between the percent elongation of such standardized round bars and the percent elongation measured on sheet specimens or wires.

True Stress and Strain

If the tensile tests are used to predict how the material will behave under other forms of loading, true stresstrue strain curves are useful. The true stress, σ, is defined as

$$\sigma = F/A, \tag{3.4}$$

where A is the instantaneous cross-sectional area corresponding to the force, F. Before necking begins, the true strain, ε, is given by

$$\varepsilon = \ln(L/L_o). \tag{3.5}$$

The engineering stress, s, is defined as the force divided by the original area, $s = F/A_o$, and the engineering strain, e, as the change in length divided by the original length, $e = \Delta L/L_o$. As long as the deformation is uniform along the gauge length, the true stress and true strain can be calculated from the engineering quantities. With constant volume, $LA = L_o A_o$,

$$A_o/A = L/L_o, \tag{3.6}$$

so $A_o/A = 1 + e$. Rewriting Equation 3.4 as $\sigma = (F/A_o)(A_o/A)$ and substituting $A_o/A = 1 + e$ and $s = F/A_o$,

$$\sigma = s(1 + e) \tag{3.7}$$

Substitution of $L/L_o = 1 + e$ into Equation 3.5 gives

$$\varepsilon = \ln(1 + e). \tag{3.8}$$

These expressions are valid only if the deformation is uniformly distributed along the gauge section. After necking starts, Equation 3.4 is still valid for true stress but the cross-sectional area at the base of the neck must be measured independently. Equation 3.5 could still be used if L and L_o were known for a gauge section centered on the middle of the neck and so short that the variations of area along its length are negligible. Then Equation 3.6 would be valid over such a gauge section, so the true strain can be calculated as

$$\varepsilon = \ln(A_o/A), \tag{3.9}$$

where A is the area at the base of the neck. Figure 3.9 shows the comparison of engineering and true stress-strain curves for the same material.

EXAMPLE PROBLEM #3.1: In a tensile test, a material fractured before necking. The true stress and strain at fracture were 630 MPa and 0.18, respectively. What is the tensile strength of the material?

Solution: The engineering strain at fracture was $e = \exp(0.18) - 1 = 0.197$. Because $s = \sigma/(1 + e)$, the tensile strength $= 630/1.197 = 526$ MPa.

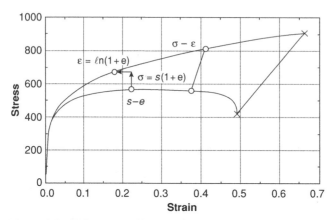

Figure 3.9. Comparison of engineering and true stress-strain curves. Before necking, a point on the true stress-strain curve ($\sigma - \varepsilon$) can be constructed from a point on the engineering stress-strain curve ($s - e$) with Equations 3.7 and 3.8. After necking, the cross-sectional area at the neck must be measured to find the true stress and strain.

Temperature Rise

Most of the mechanical energy expended by the tensile machine is liberated as heat in the tensile specimen. If the testing is rapid, little heat is lost to the surroundings and the temperature rise can be surprisingly high. With very slow testing, most of the heat is dissipated to the surroundings so the temperature rise is much less.

> **EXAMPLE PROBLEM #3.2:** Calculate the temperature rise in a tension test of a low-carbon steel after a tensile elongation of 22%. Assume that 95% of the energy goes to heat and remains in the specimen. Also assume that Figure 3.3 represents the stress strain curve of the material. For steel, the heat capacity is 447 J/kg-K and the density is 7.88×10^6 g/m^3.
>
> *Solution:* The heat released equals $0.95 \int s\,de$. From Figure 3.3, $\int s\,de = s_{\text{av}}e$ equal to about 34×0.22 ksi $= 51.6$ MPa $= 61.6$ MJ/m^3. $\Delta T = Q/C$ were $Q = 0.95(61.6 \times 10^6 \text{J/m}^3)$ and C is heat capacity per volume $= (78 \times 10^3 \text{kg/m}^3)(447 \text{ J/kg/K})$. Substituting, $\Delta T = 0.95(61.6 \times 10^6 \text{J/m}^3)/[(7.8 \times 10^3 \text{ kg/m}^3)(447 \text{ J/kg/K})] = 17°$C. This is a moderate temperature rise. For high-strength materials, the rise can be much greater.

Compression Test

Because necking limits the uniform elongation in tension, tension tests are not useful for studying the plastic stress-strain relationships at high strains. Much greater strains can be reached in compression, torsion, and bulge tests. The results from these tests can be used, together with the theory of plasticity (Chapter 5), to predict the stress-strain behavior under other forms of loading.

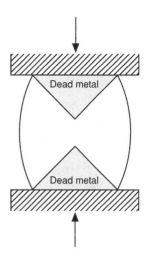

Figure 3.10. Unless the ends of a compression specimen are well lubricated, there will be a conical region of undeforming material (dead metal) at each end of the specimen. As a consequence, the mid-section will bulge out or *barrel*.

Much greater strains are achievable in compression tests than in tensile tests. However, two problems limit the usefulness of compression tests: friction and buckling. Friction on the ends of the specimen tends to suppress the lateral spreading of material near the ends (Figure 3.10). A cone-shaped region of *dead metal* (undeforming material) can form at each end with the result that the specimen becomes barrel-shaped. Friction can be reduced by lubrication, and the effect of friction can be lessened by increasing the height-to-diameter ratio, h/d, of the specimen.

If the coefficient of friction, μ, between the specimen and platens is constant, the average pressure to cause deformation is

$$P_{av} = Y(1 + (\mu d/h)/3 + (\mu d/h)^2/12 + \cdots), \qquad (3.12)$$

where Y is the true flow stress of the material. On the other hand, if there is a constant shear stress at the interface, such as would be obtained by inserting a thin film of a soft material (e.g., lead, polyethylene, or Teflon), the average pressure is

$$P_{av} = Y + (1/3)k(d/h), \qquad (3.13)$$

where k is the shear strength of the soft material. However, these equations usually do not accurately describe the effect of friction because neither the coefficient of friction nor the interface shear stress is constant. Friction is usually greatest at the edges where liquid lubricants are lost and thin films may be cut during the test by sharp edges of the specimens. Severe barreling caused by friction may cause the sidewalls to fold up and become part of the ends, as

Figure 3.11. Photograph of the end of a compression specimen. The darker central region was the original end. The lighter region outside was originally part of the cylindrical wall that folded up with the severe barreling. From G. W. Pearsall and W. A. Backofen, *Journal of Engineering for Industry, Trans ASME* v. 85B (1963) pp. 68–76.

shown in Figure 3.11. Periodic unloading to replace or relubricate the film will help reduce these effects.

Although increasing h/d reduces the effect of friction, the specimen will buckle if it is too long and slender. Buckling is likely if the height-to-diameter ratio is greater than about 3. If the test is so well lubricated that the ends of the specimen can slide relative to the platens, buckling can occur for h/d greater than equal to 1.5 (Figure 3.12).

One way to circumvent the effects of friction is to test specimens with different diameter/height ratios. The strains at several levels of stress are plotted against d/h. By the extrapolating the stresses to $d/h = 0$, the stress levels can be found for an infinitely long specimen in which the friction effects would be negligible (Figure 3.13).

During compression, the load-carrying cross-sectional area increases. Therefore, in contrast to the tension test the absolute value of engineering stress is greater than the true stress (Figure 3.14). The area increase, together

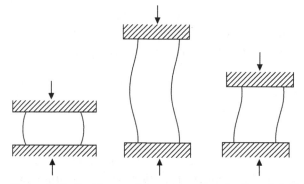

Figure 3.12. Problems with compression testing. (a) Friction at the ends prevents spreading, which results in barreling; (b) buckling of poorly lubricated specimens can occur if the height-to-diameter ratio, h/d, exceeds about 3. Without any friction at the ends (c), buckling can occur if h/d is greater than about 1.5.

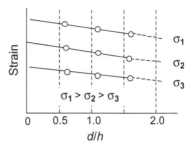

Figure 3.13. Extrapolation scheme for eliminating frictional effects in compression testing. Strains at different levels of stress (σ_1, σ_2, σ_3) are plotted for specimens of differing heights. The strain for "frictionless" conditions is obtained by extrapolating d/h to 0.

with work hardening, can lead to very high forces during compression tests unless the specimens are very small.

The shape of the engineering stress-strain curve in compression can be predicted from the true stress-strain curve in tension, assuming that absolute values of true stress in tension and compression are the same at the same absolute strain-values. Equations 3.7 and 3.8 apply, but it must be remembered that both the stress and strain are negative in compression,

$$e_{\text{comp}} = \exp(\varepsilon) - 1, \tag{3.14}$$

and

$$s_{\text{comp}} = \sigma/(1 + e). \tag{3.15}$$

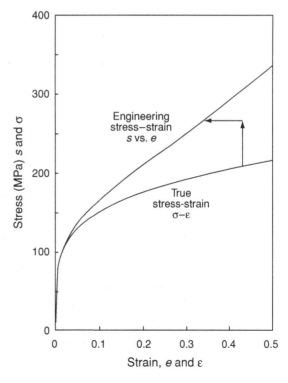

Figure 3.14. Stress-strain relations in compression for a ductile material. Each point, σ, ε, on the true stress-true strain curve corresponds to a point, s, e, on the engineering stress-strain curve. The arrows connect these points.

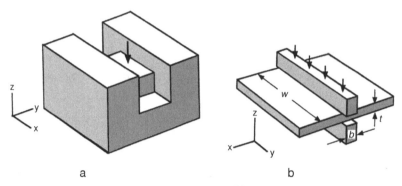

a   b

Figure 3.15. Plane-strain compression tests. (a) Compression in a channel with the side walls preventing spreading, and (b) plane-strain compression of a wide sheet with a narrow indenter. Lateral constraint forcing $\varepsilon_y = 0$ is provided by the adjacent material that is not under the indenter

EXAMPLE PROBLEM #3.3: In a tensile test, the engineering stress $s = 100$ MPa at an engineering strain of $e = 0.20$. Find the corresponding values of σ and ε. At what engineering stress and strain in compression would the values of $|\sigma|$ and $|\varepsilon|$ equal those values of σ and ε?

Solution: In the tensile test, $\sigma = s(1 + e) = 100(1.2) = 120$ MPa, $\varepsilon = \ln(1 + e) = \ln(1.2) = 0.182$. At a true strain of -0.182 in compression, the engineering strain would be $e_{\text{comp}} = \exp(-0.18) - 1 = -0.1667$ and the engineering stress would be $s_{\text{comp}} = \sigma/(1 + e) = -120$ MPa$/(1 - 0.1667) = -144$ MPa.

Compression failures of brittle materials occur by shear fractures on planes at 45° to the compression axis. In materials of high ductility, cracks may occur on the barreled surface, either at 45° to the compression axis or perpendicular to the hoop direction. In the latter case, secondary tensile stresses are responsible. These occur because the frictional constraint on the ends causes the sidewalls to bow outward. Because of this barreling, the axial compressive stress in the bowed walls is lower than in the center. Therefore, a hoop-direction tension must develop to aid in the circumferential expansion.

Plane-Strain Compression and Tension

There are two simple ways of making plane-strain compression tests. Small samples can be compressed in a channel that prevents spreading (Figure 3.15a). In this case, there is friction on the sidewalls of the channel as well as on the platens, so the effect of friction is even greater than in uniaxial compression. An alternative way of producing plane-strain compression is to use a specimen that is very wide relative to the breadth of the indenter (Figure 3.15b). This eliminates the sidewall friction but the deformation at and

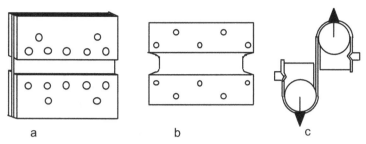

Figure 3.16. Several ways of making plane-strain tension tests on sheet specimens. All have a gauge section that is very short relative to its width: (a) Enlarged grips produced by welding to additional material, (b) reduced gage section cut into edge (c) very short gage section achieved by friction on the cylindrical grips.

near the edges deviates from plane strain. This departure from plane strain extends inward for a distance approximately equal to the indenter width. To minimize this effect, it is recommended that the ratio of the specimen width to indenter width, w/b, be about 8:1. It is also recommended that the ratio of the indenter width to sheet thickness, b/t, be about 2:1. Increasing b/t increases the effect of friction. Both of these tests simulate the plastic conditions that prevail during flat rolling of sheet and plate. They find their greatest usefulness in exploring the plastic anisotropy of materials.

Plane-strain can be achieved in tension with specimens having gauge sections that are very much wider than they are long. Figure 3.16 shows several possible specimens and specimen gripping arrangements. Such tests avoid all the frictional complications of plane-strain compression. However, the regions near the edges lack the constraint necessary to impose plane strain. At the very edge, the stress preventing contraction disappears so the stress state is uniaxial tension. Corrections must be made for departure from plane strain flow near the edges.

Biaxial Tension (Hydraulic Bulge Test)

Much greater strains can be reached in bulge tests than in uniaxial tension tests. This allows evaluation of the stress-strain relationships at high strains. A setup for bulge testing is sketched in Figure 3.17. A sheet specimen is placed

Figure 3.17. Schematic of a hydraulic bulge test of a sheet specimen. Hydraulic pressure causes biaxial stretching of the clamped sheet.

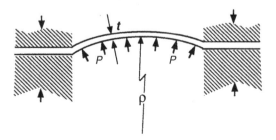

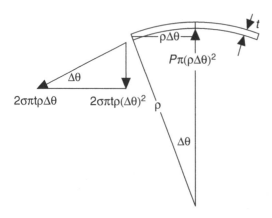

Figure 3.18. Force balance on a small circular region near the dome. The force acting upward is the pressure multiplied by the circular area. The total tangential force equals the thickness multiplied by the tangential stress. The downward force is the vertical component of this.

over a circular hole, clamped, and bulged outward by oil pressure acting on one side.

Consider a force balance on a small circular element of radius ρ near the pole when $\Delta\theta$ is small (Figure 3.18). Using the small angle approximation, the radius of this element is $\rho\Delta\theta$, where ρ is the radius of curvature. The stress, σ, on this circular region acts on an area $2\pi\rho\Delta\theta t$ and creates a tangential force equal to $2\pi\sigma\rho\Delta\theta t$. The vertical component of the tangential force is $2\pi\sigma\rho\Delta\theta t$ multiplied by $\Delta\theta$, or $2\pi\sigma\rho(\Delta\theta)^2 t$. This is balanced by the pressure, P, acting on an area $\pi(\rho\Delta\theta)^2$ and creating an upward force of $P\pi(\rho\Delta\theta)^2$. Equating the vertical forces,

$$\sigma = Pr/(2t). \tag{3.16}$$

To find the stress, the pressure, the radius of curvature, and the radial strain must be measured simultaneously. The thickness, t, is then deduced from the original thickness, t_o, and the radial strain, ε_r,

$$t = t_o\exp(-2\varepsilon_r). \tag{3.17}$$

EXAMPLE PROBLEM #3.4: Assume that in a hydraulic bulge test, the bulged surface is a portion of a sphere and that at every point on the thickness of the bulged surface the thickness is the same. (Note: This is not strictly true.) Express the radius of curvature, ρ, in terms of the die radius, r, and

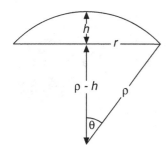

Figure 3.19. Geometry of a bulged surface.

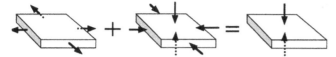

Figure 3.20. Equivalence of biaxial tension and through-thickness compression.

the bulged height, h. Also express the thickness strain at the dome in terms of r and h (see Figure 3.19).

Solution: Using the Pythagorean theorem, $(\rho - h)^2 + r^2 = \rho^2$, so $2\rho h = h^2 + r^2$ or $\rho = (r^2 + h^2)/(2h)$ where ρ is radius of the sphere and θ is the internal angle.

The area of the curved surface of a spherical segment is $A = 2\pi eh$. The original area of the circle was $A_o = \pi r^2$, so the average thickness strain is $\varepsilon_t \approx \ln[(2\pi rh)/(2hr)] = \ln(2h/r)$.

Hydrostatic compression superimposed on the state of biaxial tension at the dome of a bulge is equivalent to a state of through-thickness compression, as shown schematically in Figure 3.20. The hydrostatic part of the stress state has no effect on plastic flow. Therefore, the plastic stress-strain behavior of biaxial tension (in the plane of the sheet) and through-thickness compression are equivalent.

Torsion Test

Very large strains can be reached in torsion. The specimen shape remains constant, so there is no necking instability or barreling. There is no friction on the gage section. Therefore, torsion testing can be used to study plastic stress-strain relations to large strains. In a torsion test, each element of the material deforms in pure shear, as shown in Figure 3.21. The shear strain, γ,

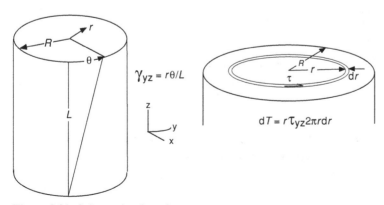

Figure 3.21. Schematic of test?

in an element is given by

$$\gamma = r\theta/L, \tag{3.18}$$

where r is the radial position of the element, θ is the twist angle, and L is the specimen length. The shear stress, τ, cannot be measured directly or even determined unequivocally from the torque. This is because the shear stress, τ, depends on γ, which varies with radial position. Therefore, τ depends on r. Consider an annular element of radius r and width dr having an area $2\pi r dr$. The contribution of this element to the total torque, T, is the product of the shear force on it, $\tau \cdot 2\pi r dr$, and the lever arm, r,

$$dT = 2\pi \tau r^2 dr \quad \text{and}$$

$$T = 2\pi \int_0^R \tau r^2 dr \tag{3.19}$$

Equation 3.19 cannot be integrated directly because τ depends on r. Integration requires substitution of the stress-strain ($\tau - \gamma$) relation. Handbook equations for torque are usually based on assuming elasticity. In this case, $\tau = G\gamma$. Substituting this and Equation 3.18 into Equation 3.19,

$$T = 2\pi(\theta/L) \int_0^R r^3 dr = (\pi/2)(\theta/L)Gr^4. \tag{3.20}$$

Because $\tau_{yz} = G\gamma_{yz}$ and $\gamma_{yz} = r\theta/L$, the shear stress varies linearly with the radial position and can be expressed as $\tau_{yz} = \tau_s(r/R)$, where τ_s is the shear stress at the surface. The value of τ_s for elastic deformation can be found from the measured torque by substituting $\tau_{yz} = \tau_s(r/R)$ into Equation 3.20,

$$T = 2\pi \int_0^r \tau_s(r/R)r^2 dr = (\pi/2)\tau_s R^3, \quad \text{or } \tau_s = 2T/(\pi R^3). \tag{3.21}$$

If the bar is not elastic, Hooke's law cannot be assumed. The other extreme is when the entire bar is plastic and the material does not work-harden. In this case, τ is a constant.

EXAMPLE PROBLEM #3.5: Consider a torsion test in which the twist is great enough that the entire cross section is plastic. Assume that the shear yield stress, τ, is a constant. Express the torque, T, in terms of the bar radius, R, and τ.

Solution: The shear force on a differential annular element at a radius r is $\tau \cdot 2\pi r \, dr$. This causes a differential torque of $dT = r(\tau \cdot 2\pi r \, dr)$. Integrating,

$$= 2\pi\tau \int\limits_0^R r^2 dr = (2/3)\pi R^3 \tau. \qquad (3.22)$$

If the torsion test is being used to determine the stress-strain relationship, the form of the stress-strain relationship cannot be assumed, so one does not know how the stress varies with radial position. One way around this problem might be to test a thin-walled tube in which the variation of stress and strain across the wall would be small enough that the variation of τ with r could be neglected. In this case, the integral (Equation 3.22) could be approximated as

$$T = 2\pi r^2 \Delta r \tau, \qquad (3.23)$$

where Δr is the wall thickness. However thin-walled tubes tend to buckle and collapse when subjected to torsion. The buckling problem can be circumvented by making separate torsion tests on two bars of slightly different diameter. The difference between the two curves is the torque-twist curve for a cylinder whose wall thickness is half the diameter difference.

The advantage of torsion tests is that very large strains can be reached, even at elevated temperatures. Because of this, torsion tests have been used to simulate the deformation in metal during hot rolling so that the effects of simultaneous hot deformation and recrystallization can be studied. It should be realized that in a torsion test, the material rotates relative to the principal stress axes. Because of this, the strain path in the material is constantly changing.

Bend Tests

Bend tests are used chiefly for materials that are very brittle and difficult to machine into tensile bars. In bending, as in torsion, the stress and strain vary with location. The engineering strain, e, varies linearly with distance, z, from the neutral plane,

$$e = z/\rho, \qquad (3.24)$$

where ρ is the radius of curvature at the neutral plane, as shown in Figure 3.22. Consider bending a plate of width w. The bending moment, dM, caused by the stress on a differential element at z is the force $\sigma w \, dz$ multiplied by the lever

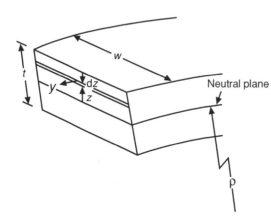

Figure 3.22. Bending moment caused by a differential element of width, w, thickness, dz, at a distance, z, from the neutral plane.

arm, z, $dM = \sigma w z dz$. The total bending moment is twice the integral of dM from the neutral plane to the outside surface,

$$M = 2 \int_0^{t/2} \sigma w z dz. \tag{3.25}$$

To integrate, σ must be expressed in terms of z. If the entire section is elastic and $w \ll t$, $\sigma = eE$ (or if $w \gg t$, $\sigma = eE/(1 - v^2)$). Substituting for e, $\sigma = (z/\rho)E$. Integrating $M = 2w(E/\rho) \int_0^{t/2} z^2 dz$,

$$M = 2w(E/\rho)(t/2)^3/3 = w(E/\rho)t^3/12. \tag{3.26}$$

The surface stress is $\sigma_s = E(t/2)/\rho$. Substituting $\rho = wEt^3/(12M)$ from Equation 3.26,

$$\sigma_s = 6M/(wt^2). \tag{3.27}$$

In three-point bending, $M = FL/4$, and in four-point bending, $M = FL/2$, as shown in Figure 3.23. The relation between the moment and the stress is different if plastic deformation occurs.

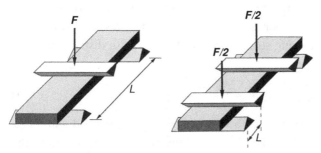

Figure 3.23. Three-point bending test (left) and four-point bending (right). Note that L is defined differently for each test.

EXAMPLE PROBLEM #3.6: Consider a four-point bend test on a flat sheet of width, w, of a material for which the stress to cause plastic deformation is a constant, Y. Assume also that the bend is sharp enough so that the entire thickness is plastic. Derive an expression that can be used to determine Y from the force, F.

Solution: The differential force on an element of thickness, dz, at a distance z from the center is $Y(w dz)$. This force causes a differential moment, $dM = wYzdz$. The net moment is twice the integral of dM from the center to the surface $(z = t/2)$. Then $M = 2wY(t/2)^2/2 = wt^2Y/4$. Substituting $M = FL/2$ for four-point bending, $Y = 2FL/wt^2$.

Measurements of fracture strengths of brittle materials are usually characterized by a large amount of scatter because of pre-existing flaws, so many duplicate tests are usually required. The fracture stress is taken as the value of the surface stress, σ_s, at fracture. This assumes that no plastic deformation has occurred (see Chapters 11 and 12).

Hardness Tests

Hardness tests are very simple to conduct, and they can be performed on production parts as quality control checks without destroying the part. They depend on measuring the amount of deformation caused when a hard indenter is pressed into the surface with a fixed force. The disadvantage is that although hardness of a material depends on the plastic properties, the stress-strain relation cannot be obtained. Figure 3.24 shows the indenters used for various tests. The Rockwell tests involve measuring the depth of indentation. There are several different Rockwell scales, each of which uses different shapes and sizes of indenters and different loads. Conversion from one scale to another is approximate and empirical.

Brinell hardness is determined by pressing a ball, 1 cm in diameter, into the surface under a fixed load (500 or 3000 kg). The diameter of the impression is measured with an eyepiece and converted to hardness. The scale is such that the Brinell hardness number, H_B, is given by

$$H_B = F/A_s, \tag{3.25}$$

where F is the force expressed in kg and A_s is the spherical surface area of the impression in mm^2. This area can be calculated from

$$A_s = \pi D[D - (D^2 - d^2)^{1/2}]/2, \tag{3.26}$$

where D is the ball diameter and d is the diameter of the impression, but it is more commonly found from tables.

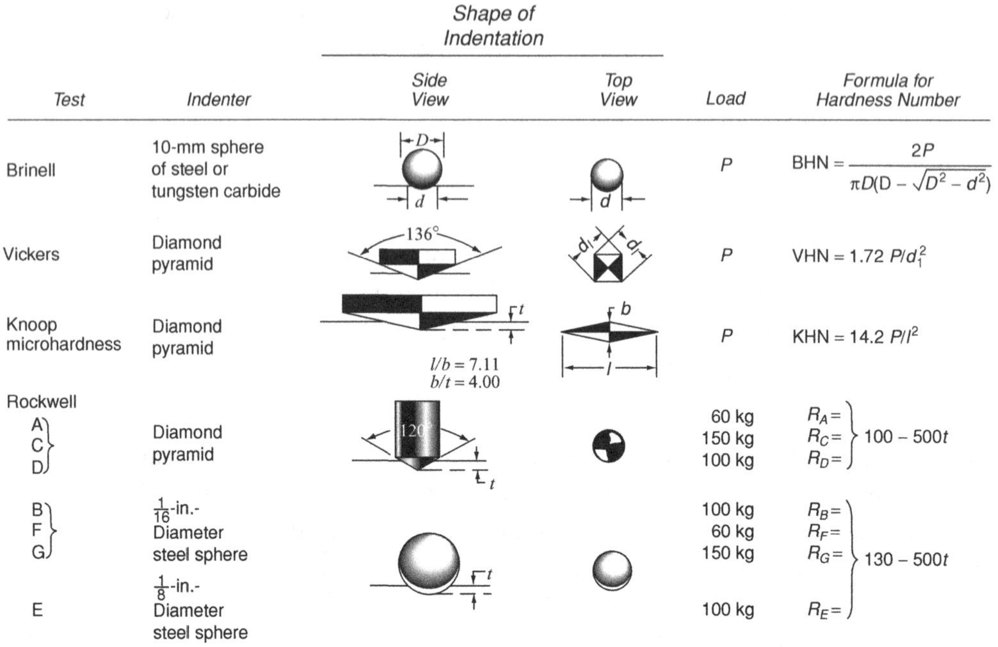

Test	Indenter	Shape of Indentation		Load	Formula for Hardness Number
		Side View	Top View		
Brinell	10-mm sphere of steel or tungsten carbide			P	$BHN = \dfrac{2P}{\pi D(D - \sqrt{D^2 - d^2})}$
Vickers	Diamond pyramid	136°		P	$VHN = 1.72 \, P/d_1^2$
Knoop microhardness	Diamond pyramid	$l/b = 7.11$ $b/t = 4.00$		P	$KHN = 14.2 \, P/l^2$
Rockwell A) C} D)	Diamond pyramid	120°		60 kg 150 kg 100 kg	$\left.\begin{array}{l} R_A = \\ R_C = \\ R_D = \end{array}\right\} 100 - 500t$
B) F} G)	$\frac{1}{16}$-in.- Diameter steel sphere			100 kg 60 kg 150 kg	$\left.\begin{array}{l} R_B = \\ R_F = \\ R_G = \end{array}\right\} 130 - 500t$
E	$\frac{1}{8}$-in.- Diameter steel sphere			100 kg	$R_E =$

Figure 3.24. Various hardness tests. From H. W. Hayden, William G. Moffat and John Wulff, *Structure and Properties of Materials, Vol. III Mechanical Behavior*, Wiley (1965).

Brinell data can be used to determine the Meyer hardness, H_M, which is defined as the force divided by the projected area of the indentation, $\pi d^2/4$. The Meyer hardness has greater fundamental significance because it is relatively insensitive to load. Vickers and Knoop hardnesses are also defined as the indentation load divided by the projected area. They are very nearly equal for the same material. Indentations under different loads are geometrically similar unless the indentation is so shallow that the indenter tip radius is not negligible compared to the indentation size. For this reason, the hardness does not depend on the load. For Brinell, Knoop, and Vickers hardnesses, a rule of thumb is that the hardness is about three times the yield strength when expressed in the same units. (Note that hardness is conventionally expressed in kg/mm^2 and strength in MPa. To convert kg/mm^2 to MPa, multiply by 9.807. Figure 3.25 shows approximate conversions between several hardness scales).

Mineralogists frequently classify minerals by the Moh's scratch hardness. The system is based on ranking minerals on a scale of 1 to 10. The scale is such that a mineral higher on the scale will scratch a mineral lower on the scale. The scale is arbitrary and not well-suited to metals as most metals tend to fall in the range between 4 and 8. Actual values vary somewhat with how the test

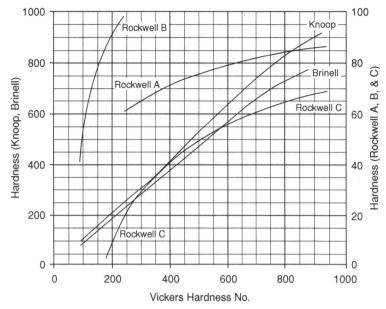

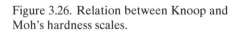

Figure 3.25. Approximate relations between several hardness scales. For Brinell (3000 kg) and Knoop, read left. For Rockwell A, B, and C scales, read right. Data from *Metals Handbook*, vol. 8, 9th ed, American Society for Metals (1985).

is made (e.g., the angle of inclination of the scratching edge). Figure 3.26 gives the approximate relationship between Moh's and Vickers hardnesses.

Hardness tests are normally made with indenters that do not deform. However, it is possible to measure hardness when the indenters do deform. This is useful at very high temperatures where it is impossible to find a suitable material for the indenter. Defining the hardness as the indentation force

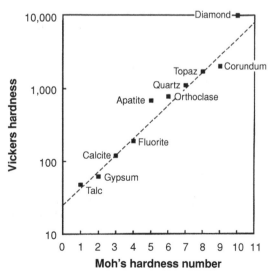

Figure 3.26. Relation between Knoop and Moh's hardness scales.

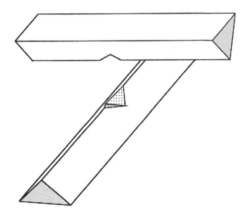

Figure 3.27. Mutual indentation of two wedges. The hardness is the force per unit area.

divided by the area of indentation, the hardness, H, of a material is proportional to its yield strength,

$$H = cY, \qquad (3.27)$$

where c depends mainly on the geometry of the test. For example, in a Brinell test where the indenting ball is very much harder than the test material, $c \approx 2.8$. For mutual indentation of cross wedges, $c \approx 3.4$, (Figure 3.27) and for crossed cylinders, $c \approx 2.4$. The constant c has been determined for other geometries and for cases where the indenters have different hardnesses. The forces in automobile accidents have been estimated from examining the depths of mutual indentations.

Notes

Leonardo da Vinci (1452–1519) described a tensile test of a wire. Fine sand was fed through a small hole into a basket attached to the lower end of the wire until the wire broke. Figure 3.28 shows his sketch of the apparatus.

The yield point effect in linear polymers may be experienced by pulling the piece of plastic sheet that holds a six-pack of carbonated beverage cans together. When one pulls hard enough, the plastic will yield and the force

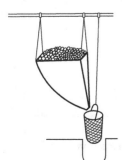

Figure 3.28. Leonardo's sketch of a system for tensile testing wire. The load was applied by pouring sand into a basket suspended from the wire.

Figure 3.29. Galileo's sketch of a bending test.

drop. A small thinned region develops. As the force is continued, this region will grow. The yield point effect in low-carbon steel may be experienced by bending a low-carbon steel wire (florist's wire works well). First, heat the wire in a flame to anneal it. Then bend a 6-in. length by holding it only at the ends. Instead of bending uniformly, the deformation localizes to form several sharp kinks. Why? Bend an annealed copper wire for comparison.

Galileo Galilel (1564–1642) stated that the strength of a bar in tension was proportional to its cross-sectional area and independent of its length. The apparatus suggested by Galileo for bending tests is illustrated in Figure 3.29.

REFERENCES

Tensile Testing, ASM International, Materials Park, OH (1992).
Percy Bridgman, *Studies in Large Plastic Flow and Fracture*, McGraw-Hill, NY (1952).
Metals Handbook, 9th ed. vol. 8, *Mechanical Testing*, ASM (1985).
D. K. Felbeck and A. G. Atkins, *Strength and Fracture of Engineering Solids*, Prentice-Hall (1984).

Problems

1. The results of a tensile test on a steel test bar are given in Table 3.1. The initial gauge length was 25.0 mm and the initial diameter was 5.00 mm. The diameter at the fracture was 2.6 mm. The engineering strains and stresses are given in Table 3.1.
   A. Plot the engineering stress-strain curve.
   B. Determine Young's modulus, the 0.2% offset yield strength, the tensile strength, the percent elongation, and the percent reduction of area.
2. Construct the true stress/true strain curve for the material in Problem 1. Note that necking starts at maximum load, so the construction should be stopped at this point.

Table 3.1. *Results of a tension test*

Strain	Stress (MPa)	Strain	Stress (MPa)	Strain	Stress (MPa)
0.0	0.0	0.06	319.8	0.32	388.4
0.0002	42.	0.08	337.9	0.34	388.0.
0.0004	83.	0.10	351.1	0.38	386.5
0.0006	125.	0.15	371.7	0.40	384.5
0.0015	155.	0.20	382.2	0.42	382.5
0.005	185.	0.22	384.7	0.44	378.
0.02	249.7	0.24	386.4	0.46	362.
0.03	274.9	0.26	387.6	0.47	250
0.04	293.5	0.28	388.3		
0.05	308.0	0.30	388.9		

3. Determine the engineering strains, e, and the true strains, ε for each of the following:
   A. Extension from $L = 1.0$ to $L = 1.1$
   B. Compression from $h = 1$ to $h = 0.9$
   C. Extension from $L = 1$ to $L = 2$
   D. Compression from $h = 1$ to $h = 1/2$.

4. The ASM *Metals Handbook* (Vol. 1, 8th ed., p 1008) gives the percent elongation in a 2-in. gauge section for annealed electrolytic tough-pitch copper as
   • 55% for a 0.505 in. diameter bar
   • 45% for a 0.030 in. thick sheet and
   • 38.5% for a 0.010 in diameter wire.
   Suggest a reason for the differences.

5. The area under an engineering stress-strain curve up to fracture is the energy per unit volume. The area under a true stress-strain curve up to fracture is also the energy per unit volume. If the specimen necks, these two areas are not equal. What is the difference? Explain.

6. Suppose it is impossible to use an extensometer on the gauge section of a test specimen. Instead, a button-head specimen (Figure 3.2c) is used and the strain is computed from the cross-head movement. There are two possible sources of error with this procedure. One is that the gripping system may deform elastically and the other is that the button head may be drawn partly through the collar. How would each of these errors affect the calculated true stress and true strain?

7. Two strain gauges were mounted on opposite sides of a tensile specimen. Strains measured as the bar was pulled in tension were used to compute Young's modulus. Readings from one gauge gave a modulus much greater than those from the other gauge. What was the probable cause of this discrepancy?

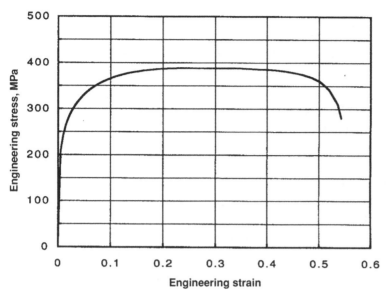

Figure 3.30. Engineering stress-strain curve for a low-carbon steel.

8. Discuss the how friction and inhomogeneous deformation affect the results from:
   A. The two types of plane-strain compression tests illustrated in Figure 3.19.
   B. The plane-strain tension tests illustrated in Figure 3.20.
9. Sketch the three-dimensional Mohr's stress and strain diagrams for a plane-strain compression test.
10. Draw a Mohr's circle diagram for the surface stresses in a torsion test, showing all three principal stresses. At what angle to the axis of the bar are the tensile stresses the greatest?
11. For a torsion test, derive equations relating the angle, ψ, between the axis of the largest principal stress and the axial direction and the angle, ψ, between the axis of the largest principal strain and the axial direction in terms of L, r, and the twist angle, θ. Note that for finite strains, these two angles are not the same.
12. The engineering stress-strain curve from a tension test on a low-carbon steel is plotted in Figure 3.30. From this construct the engineering stress-strain curve in compression, neglecting friction.
13. Derive an expression relating the torque, T, in a tension test to the shear stress at the surface, τ_s, in terms of the bar diameter, D, assuming that the bar is:
    A. Entirely elastic so τ varies linearly with the shear strain, γ.
    B. Entirely plastic and does not work-harden so the shear stress, τ, is constant.

14. The principal strains in a circular bulge test are the thickness strain, ε_t, the circumferential (hoop) strain, ε_c, and the radial strain, ε_r. Describe how the ratio, $\varepsilon_c/\varepsilon_r$, varies over the surface of the bulge. Assume that the sheet is locked at the opening.

15. Derive an expression for the fracture strength, S_f, in bending as a function of F_f, L, w, and t for the three-point bending of a specimen having a rectangular cross-section, where F_f is the force at fracture, L is the distance between supports, w is the specimen width, and t is the specimen thickness. Assume the deformation is elastic and the deflection, y, in bending is given by $y = \alpha FL^3/(EI)$, where a is a constant and E is Young's modulus. How would you expect the value of A to depend on the ratio of t/w?

16. Equation 3.24 gives the stresses at the surface of bend specimens. The derivation of this equation is based on the assumption of elastic behavior. If there is plastic deformation during the bend test, will the stress predicted by this equation be too low, too high, either too high or too low depending on where the plastic deformation occurs, or correct?

17. By convention, Brinell, Meyer, Vickers, and Knoop hardness numbers are stresses expressed in units of kg/mm^2, which is not an SI unit. To what stress, in MPa, does a Vickers hardness of 100 correspond?

18. In making Rockwell hardness tests, it is important that the bottom of the specimen is flat so that the load does not cause any bending of the specimen. On the other hand, this is not important in making Vickers or Brinell hardness tests. Explain.

4   Strain Hardening of Metals

Introduction

With elastic deformation, the strains are proportional to the stress so every level of stress causes some elastic deformation. On the other hand, a definite level of stress must be applied before any plastic deformation occurs. As the stress is further increased, the amount of deformation increases but not linearly. After plastic deformation starts, the total strain is the sum of the elastic strain (which still obeys Hooke's law) and the plastic strain. Because the elastic part of the strain is usually much less than the plastic part, it will be neglected in this chapter, and the symbol ε will signify the true plastic strain.

The terms *strain hardening* and *work hardening* are used interchangeably to describe the increase of the stress level necessary to continue plastic deformation. The term *flow stress* is used to describe the stress necessary to continue deformation at any stage of plastic strain. Mathematical descriptions of true stress-strain curves are needed in engineering analyses that involve plastic deformation such as predicting energy absorption in automobile crashes, designing of dies for stamping parts, and analyzing the stresses around cracks. Various approximations are possible. Which approximation is best depends on the material, the nature of the problem, and the need for accuracy. This chapter will consider several approximations and their applications.

Mathematical Approximations

The simplest model is one with no work-hardening. The flow stress, σ, is independent of strain, so

$$\sigma = Y, \tag{4.1}$$

where Y is the tensile yield strength (see Figure 4.1a). For linear work-hardening (Figure 4.1b),

$$\sigma = Y + A\varepsilon. \tag{4.2}$$

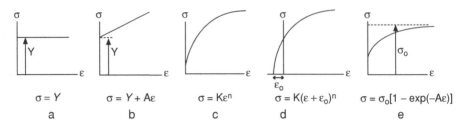

Figure 4.1. Mathematical approximations of the true stress-strain curve.

It is more common for materials to work-harden with a hardening rate that decreases with strain. For many metals, a log-log plot of true stress vs. true strain is nearly linear. In this case, a power law,

$$\sigma = K\varepsilon^n, \tag{4.3}$$

is a reasonable approximation (Figure 4.1c). A better fit is often obtained with

$$\sigma = K(\varepsilon + \varepsilon_o)^n \tag{4.4}$$

(see Figure 4.1d). This expression is useful where the material has undergone a pre-strain of ε_o.

Still another model is a saturation model suggested by Voce (Figure 4.1e) is

$$\sigma = \sigma_o[1 - \exp(-A\varepsilon)]. \tag{4.5}$$

Equation 4.5 predicts that the flow stress approaches an asymptote, σ_o, at large strains. This model seems to be reasonable for a number of aluminum alloys.

EXAMPLE PROBLEM #4.1: Two points on a true stress-strain curve are $\sigma = 222$ MPa at $\varepsilon = 0.05$ and $\sigma = 303$ MPa at $\varepsilon = 0.15$.

A. Find the values of K and n in Equation 4.3 that best fit the data. Then using these values of K and n, predict the true stress at a strain of $\varepsilon = 0.30$.
B. Find the values of σ_o and A in Equation 4.5 that best fit the data. Then using these values of K and n, predict the true stress at a strain of $\varepsilon = 0.30$.

Solution: A. Apply the power law at two levels of stress, σ_2 and σ_1. Then taking the ratios $\sigma_2/\sigma_1 = (\varepsilon_2/\varepsilon_1)^n$ and finally taking the natural logs of both sides, $n = \ln(\sigma_2/\sigma_1)/\ln(\varepsilon_2/\varepsilon_1) = \ln(303/222)/\ln(0.15/0.05) = 0.283$. $K = \sigma/\varepsilon^n = 303/0.15^{0.283} = 518.5$ MPa.

At $e = 0.30$, $\sigma = 518.4(0.30)^{0.283} = 368.8$ MPa.

B. Appling the saturation model to two levels of stress, σ_2 and σ_1, $\sigma_1 = \sigma_o[1 - \exp(-A\varepsilon_1)]$ and $\sigma_2 = \sigma_o[1 - \exp(A\varepsilon_2)]$. Then taking the ratios, $\sigma_2/\sigma_1 = [1 - \exp(-A\varepsilon_2)]/[1 - \exp(-A\varepsilon_1)]$. Substituting $\sigma_2/\sigma_1 = 303/222$

$= 1.365, \varepsilon_1 = 0.05$ and $\varepsilon_2 = 0.15$ and, solving by trial and error, $A = 25.2$.
$\sigma_o = 303[1 - \exp(-25.2 \times 0.15)] = 296$ MPa. At $e = 0.30, s = 310[1 - \exp(-25.2 \times 0.30)] = 309.8$ MPa.

Power-Law Approximation

The most commonly used expression is the simple power law (Equation 4.3). Typical values of the exponent n are in the range of 0.1 to 0.6. Table 4.1 lists K and n for various materials. As a rule, high-strength materials have lower n-values than low-strength materials. Figure 4.2 shows that the exponent, n, is a measure of the persistence of hardening. If n is low, the work-hardening rate is initially high but the rate decreases rapidly with strain. On the other hand, with a large n, the initial work-hardening is less rapid but continues to high strains. If $\sigma = K\varepsilon^n$, $\ln\sigma = \ln K + n\ln\varepsilon$ so the true stress-strain relation plots as a straight line on log-log coordinates as shown in Figure 4.3. The exponent, n, is the slope of the line. The pre-exponential, K, can be found by extrapolating to $\varepsilon = 1.0$. K is the value of σ at this point.

The level of n is particularly significant in stretch forming because it indicates the ability of a metal to distribute the strain over a wide region. Often a log-log plot of the true stress-strain curve deviates from linearity at low or high strains. In such cases, it is still convenient to use Equation 4.3 over the strain range of concern. The value of n is then taken as the slope of the linear portion of the curve,

$$n = d(\ln\sigma)/d(\ln\varepsilon) = (\varepsilon/\sigma)d\sigma/d\varepsilon. \tag{4.6}$$

Necking

As a tensile specimen is extended, the level of true stress, σ, rises but the cross-sectional area carrying the load decreases. The maximum load-carrying

Table 4.1. *Typical values of n and K**

Material	Strength coefficient, K (MPa)	Strain hardening exponent, n
low-carbon steels	525 to 575	0.20 to 0.23
HSLA steels	650 to 900	0.15 to 0.18
austenitic stainless	400 to 500	0.40 to 0.55
copper	420 to 480	0.35 to 0.50
70/30 brass	525 to 750	0.45 to 0.60
aluminum alloys	400 to 550	0.20 to 0.30

* From various sources including Hosford and Caddell, *Metal Forming; Mechanics and Metallurgy*, Prentice Hall (1983).

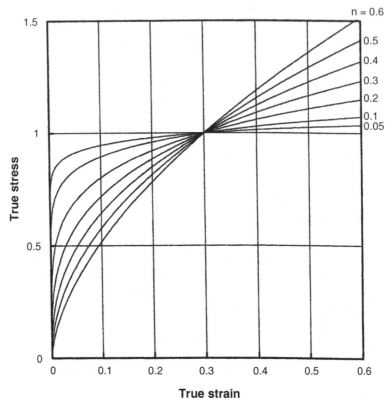

Figure 4.2. True stress–strain curves for $\sigma = K\varepsilon^n$ with several values of n. To make the effect of n on the shape of the curves apparent, the value of K for each curve has been adjusted so that it passes through $\sigma = 1$ at $\varepsilon = 0.3$.

capacity is reached when $dF = 0$. The force equals the true stress multiplied by the actual area, $F = \sigma A$. Differentiating gives

$$A\,d\sigma + \sigma\,dA = 0. \tag{4.7}$$

Because the volume, AL, is constant, $dA/A = -dL/L = -d\varepsilon$. Rearranging terms, $d\sigma = -\sigma\,dA/A = \sigma\,d\varepsilon$, or

$$d\sigma/d\varepsilon = \sigma. \tag{4.8}$$

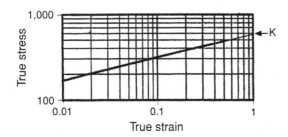

Figure 4.3. A plot of the true stress-strain curve on logarithmic scales. Because $\sigma = K\varepsilon^n$, $\ln\sigma = \ln K + n\ln\varepsilon$. The straight line indicates that $\sigma = K\varepsilon^n$ holds. The slope is equal to n, and K equals the intercept at $\varepsilon = 1$.

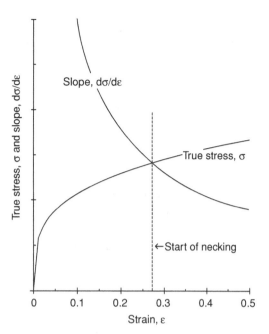

Figure 4.4. The condition for necking in a tension test is met when the true stress, σ, equals the slope, $d\sigma/d\varepsilon$, of the true stress-strain curve.

Equation 4.8 simply states that the maximum load is reached when the rate of work-hardening is numerically equal to the stress level.

As long as $d\sigma/d\varepsilon > \sigma$, deformation will occur uniformly along the test bar. However, once the maximum load is reached ($d\sigma/d\varepsilon = \sigma$), the deformation will localize. Any region that deforms even slightly more than the others will have a lower load-carrying capacity and the load will drop to that level. Other regions will cease to deform, so deformation will localize into a neck. Figure 4.4 is a graphical illustration.

If a mathematical expression is assumed for the stress-strain relationship, the limit of uniform elongation can be found analytically. For example, with power-law hardening, Equation 4.3, $\sigma = K\varepsilon^n$, and $d\sigma/d\varepsilon = nK\varepsilon^{n-1}$. Substituting into Equation 4.8 gives $K\varepsilon^n = nK\varepsilon^{n-1}$, which simplifies to

$$\varepsilon = n, \tag{4.9}$$

so the strain at the start of necking equals the strain-hardening exponent, n. Uniform elongation in a tension test occurs before necking. Therefore, materials with a large n-value have large uniform elongations.

Because the ultimate tensile strength is simply the engineering stress at maximum load, the power-law hardening rule can be used to predict it. First, find the true stress at maximum load by substituting the strain, n, into Equation 4.3,

$$\sigma_{\text{max load}} = Kn^n. \tag{4.10}$$

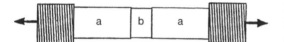

Figure 4.5. Stepped tensile specimen. The initial cross-sectional area of region b is f multiplied by the cross-sectional area of A.

Then substitute $\sigma = \sigma/(1+\varepsilon) = \sigma\exp(-\varepsilon) = \sigma\exp(-n)$ into Equation 4.10; the tensile strength is expressed as

$$\sigma_{\max} = Kn^n\exp(-n) = K(n/e)^n, \tag{4.11}$$

where e is the base of natural logarithms.

Similarly, the uniform elongation and tensile strength may be found for other approximations to the true stress-strain curve.

> **EXAMPLE PROBLEM #4.2:** A material has a true stress-strain curve that can be approximated by $\sigma = Y + A\varepsilon$. Express the uniform elongation, ε, in terms of the constants, A and Y.
>
> *Solution:* The maximum load is reached when $d\sigma/d\varepsilon = \sigma$. Equating $d\sigma/d\varepsilon = A$ and $\sigma = Y + A\varepsilon$, the maximum load is reached when $A = Y + A\varepsilon$ or $\varepsilon = 1 - Y/A$. The engineering strain is $e = \exp(\varepsilon) - 1 = \exp(1 - Y/A) - 1$.

Work per Volume

The area under the true stress-strain curve is the work per volume, w, expended in the deforming a material, that is, $w = \int\sigma d\varepsilon$. With power-law hardening,

$$w = K\varepsilon^{n+1}/(n+1). \tag{4.12}$$

This is sometimes called the *tensile toughness*.

Localization of Strain at Defects

If the stresses that cause deformation in a body are not uniform, the deformation will be greatest where the stress is highest and least where the stress is lowest. The differences between the strains in the different regions depend on the value of n. If n is large, the difference will be less than if n is low. For example, consider a tensile bar in which the cross-sectional area of one region is a little less than in the rest of the bar (Figure 4.5). Let these areas be A_{ao} and A_{bo}.

The tensile force, F, is the same in both regions, $F_a = F_b$, so

$$\sigma_a A_a = \sigma_b A_b. \tag{4.13}$$

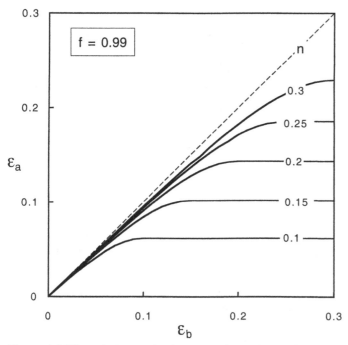

Figure 4.6. The relative strains in two regions of a tensile bar having different initial cross-sectional areas.

Because $\varepsilon = \ln(L/L_o)$ and $(L/L_o) = (A_o/A)$, the instantaneous area, A, may be expressed as $A = A_o \exp(-\varepsilon)$, so $A_a = A_{ao}\exp(-\varepsilon_a)$, $A_b = A_{bo}\exp(-\varepsilon_b)$. Assuming power-law hardening (Equation 5.3), $\sigma_a = K\varepsilon_a^n$, and $\sigma_b = K\varepsilon_b^n$. Substituting into Equation 4.13,

$$K\varepsilon_a^n A_{ao}\exp(-\varepsilon_a) = K\varepsilon_b^n A_{bo}\exp(-\varepsilon_b). \qquad (4.14)$$

Now simplify by substituting $f = A_{bo}/A_{ao}$,

$$\varepsilon_a^n\exp(-\varepsilon_a) = f\varepsilon_b^n\exp(-\varepsilon_b). \qquad (4.15)$$

Equation 4.15 can be numerically evaluated to find ε_a as a function of ε_b. Figure 4.6 shows that with low values of n, the region with the larger cross-section deforms very little. On the other hand, if n is large, there is appreciable deformation in the thicker region, so more overall stretching will have occurred when the thinner region fails. This leads to greater formability.

EXAMPLE PROBLEM #4.3: To ensure that the neck in a tensile bar will occur at the middle of the gauge section, the machinist made the bar with a 0.500-in. diameter in the middle of the gauge section and machined the rest of it to a diameter of 0.505 in. After testing, it was found that the diameter away from the neck was 0.470 in. Assume that the stress-strain relation follows the power law, Equation 5.3. What is the value of n?

Solution: Let the regions with the larger and smaller diameters be designated a and b. When the maximum load is reached, the strain in b is $\varepsilon_b = n$ and the strain in region a is $\varepsilon_a = \ln(A_o/A) = \ln(d_o/d)^2 = 2\ln(0.505/0.470) = 0.1473$. $f = (0.500/0.505)^2 = 0.9803$. Substituting $\varepsilon_b = n$, $e_a = 0.1473$ and $f = 0.98$ into Equation 5.15, $0.1437^n \exp(-0.1437) = 0.9803n^n \exp(-n)$. Solving by trial and error, $n = 0.238$. Note that there would have been a large error if the effect of the different initial cross-sections had been ignored and n had been taken as $2\ln(0.505/0.470) = 0.1473$.

Notes

Professor Zdzislaw Marciniak of the Technical University of Warsaw was the first to analyze the effect of a small defect or inhomogeneity on the localization of plastic flow.[*] His analysis of necking forms the basis for understanding superplastic elongations (Chapter 6) and for calculating forming limit diagrams (Chapter 22). He was the Acting Rector, Senior Professor, and Director of the Institute of Metal Forming at the Technical University of Warsaw and has published a number of books in Polish on metal forming and many papers in the international literature.

One can experience work-hardening by holding the ends of a wire in one's fingers and bending it, straightening it, and bending it again. The second time the bend will occur in a different spot than the first bend. Why?

Problems

1. What are the values of K and n in Figure 4.3?
2. True stress-true strain data from a tension test are given in Table 4.2:
   A. Plot true stress vs. true strain on a logarithmic plot.
   B. What does your plot suggest about n in Equation 4.3?
   C. What does your plot suggest about a better approximation?
3. The true stress-strain curve of a material obeys the power hardening law with $n = 0.18$. A piece of this material was given a tensile strain of $\varepsilon = 0.03$

Table 4.2. *Results of a tension test*

Strain	Stress (MPa)	Strain	Stress (MPa)
0.00	0.00	0.10	250.7
0.01	188.8	0.15	270.6
0.02	199.9	0.20	286.5
0.05	223.5		

[*] Z. Marciniak and K. Kuczynski, *Int. J. Mech. Sci.* v. 9, (1967) p. 609.

before being sent to a laboratory for tension testing. The lab workers were unaware of the pre-strain and tried to fit their data to Equation 4.3.

A. What value of n would they report if they determined n from the elongation at maximum load?

B. What value of n would they report if they determined n from the loads at $\varepsilon = 0.05$ and 0.15?

4. In a tension test, the following values of engineering stress and strain were found: $s = 133.3$ MPa at $e = 0.05$, $s = 155.2$ MPa at $e = 0.10$ and $s = 166.3$ MPa at $e = 0.15$.

A. Determine whether the data fit Equation 4.3.

B. Predict the strain at necking.

5. Two points on a stress-strain curve for a material are $\sigma = 278$ MPa at $\varepsilon = 0.08$ and $\sigma = 322$ MPa at $\varepsilon = 0.16$.

A. Find K and n in the power-law approximation and predict σ at $\varepsilon = 0.20$.

B. For the approximation $\sigma = K(\varepsilon_o + \varepsilon)^n$ (Equation 4.4) with $\varepsilon_o = 0.01$, find K and n and predict ε at $\varepsilon = 0.20$.

6. The tensile stress–strain curve of a certain material is best represented by a saturation model, $\sigma = \sigma_o[1 - \exp(-A\varepsilon)]$.

A. Derive an expression for the true strain at maximum load in terms of the constants A and σ_o.

B. In a tension test, the maximum load occurred at an engineering strain of $e = 21\%$, and the tensile strength was 350 MPa. Determine the values of the constants A and σ_o for the material. (Remember that the tensile strength is the maximum engineering stress.)

7. A material has a stress-strain relation that can be approximated by $\sigma = 150 + 185\varepsilon$.

A. What percent uniform elongation should be expected in a tension test?

B. What is the material's tensile strength?

8. A. Derive expressions for the true strain at the onset of necking if the stress-strain curve is given by $\sigma = K(\varepsilon_o + \varepsilon)^n$ (Equation 4.4).

B. Write an expression for the tensile strength.

9. Consider a tensile specimen made from a material that obeys the power-hardening law with $K = 400$ MPa and $n = 0.20$. Assume K is not sensitive to strain rate. One part of the gauge section has an initial cross-sectional area that is 0.99 times that of the rest of the gauge section. What will be the true strain in the larger area after the smaller area necks and reduces to 50% of its original area?

10. Consider a tensile bar that was machined so that most of the gauge section was 1.00 cm in diameter. One short region in the gauge section has a diameter 0.5% smaller (0.995 cm). Assume the stress-strain curve of the material is described by the power law with $K = 330$ MPa and $n = 0.23$, and the flow stress is not sensitive to the strain rate. The bar was

pulled in tension well beyond maximum load and it necked in the reduced section.

A. Calculate the diameter away from the reduced section.

B. Suppose that an investigator had not known that the bar initially had a reduced section and had assumed that the bar had a uniform initial diameter of 1.00 cm. Suppose that she measured the diameter away from the neck and had used that to calculate n. What value of n would she have calculated?

11. A tensile bar was machined so that most of the gauge section had a diameter of 0.500 cm. One small part of the gauge section had a diameter 1% smaller (0.495 cm). Assume power-law hardening with $n = 0.17$. The bar was pulled until necking occurred.

A. Calculate the uniform elongation (percent) away from the neck.

B. Compare this with the uniform elongation that would have been found if there were no initial reduced section.

12. Repeated cycles of freezing of water and thawing of ice will cause copper pipes to burst. Water expands about 8.3% when it freezes.

A. Consider a copper tube as a capped cylinder that cannot lengthen or shorten. If it were filled with water, what would be the circumferential strain in the wall when the water freezes?

B. How many cycles of freezing/thawing would it take to cause the tube walls to neck? Assume the tube is filled after each thawing. Assume $n = 0.55$.

Plasticity Theory

Introduction

Plasticity theory deals with yielding of materials under complex states of stress. A yield criteria allows one to decide whether or not a material will yield under a stress state. Flow rules predict the shape changes that will occur if it does yield. With plasticity theory, tensile test data can be used to predict the work-hardening during deformation under such complex stress states. These relations are a vital part of computer codes for predicting crashworthiness of automobiles and codes for designing forming dies.

Yield Criteria

The concern here is to describe mathematically the conditions for yielding under complex stresses. A *yield criterion* is a mathematical expression of the stress states that will cause yielding or plastic flow. The most general form of a yield criterion is

$$f(\sigma_x, \sigma_y, \sigma_z, \tau_{yz}, \tau_{zx}, \tau_{xy}) = C, \tag{5.1}$$

where C is a material constant. For an isotropic material this can be expressed in terms of principal stresses,

$$f(\sigma_1, \sigma_2, \sigma_3) = C. \tag{5.2}$$

The yielding of most solids is independent of the sign of the stress state. Reversing the signs of all the stresses has no effect on whether a material yields. This is consistent with the observation that for most materials, the yield strengths in tension and compression are equal.* Also, for most solid materials

---

* This may not be true when the loading path is changed during deformation. Directional differences in yielding behavior after prior straining are called a *Bauschinger effect*. It is also not true if mechanical twinning is an important deformation mechanism. The compressive yield strengths of polymers are generally greater than the tensile yield strengths (see Chapter 12).

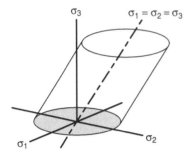

Figure 5.1. A yield locus is the surface of a body in three-dimensional stress space. Stress states on the locus will cause yielding. Those inside the locus will not cause yielding.

yielding is independent of the level of mean normal stress, σ_m,

$$\sigma_m = (\sigma_1 + \sigma_2 + \sigma_3)/3. \tag{5.3}$$

It will be shown later that this is equivalent to assuming that plastic deformation causes no volume change. This assumption of constancy of volume is certainly reasonable for crystalline materials that deform by slip and twinning because these mechanisms involve only shear. With slip and twinning, only the shear stresses are important. With this simplification, the yield criteria must be of the form

$$f[(\sigma_2 - \sigma_3),(\sigma_3 - \sigma_1),(\sigma_1 - \sigma_2)] = C. \tag{5.4}$$

In terms of the Mohr's stress circle diagrams, only the sizes of the circles (not their positions) are of importance in determining whether yielding will occur. In three-dimensional stress space (σ_1 vs. σ_2 vs. σ_3), the locus can be represented by a cylinder parallel to the line $\sigma_1 = \sigma_2 = \sigma_3$, as shown in Figure 5.1.

Tresca (maximum shear stress criterion)

The simplest yield criterion is one first proposed by Tresca. It states that yielding will occur when the largest shear stress reaches a critical value. The largest shear stress is $\tau_{max} = (\sigma_{max} - \sigma_{min})/2$, so the Tresca criterion can be expressed as

$$\sigma_{max} - \sigma_{min} = C. \tag{5.5}$$

If the convention is maintained that $\sigma_1 \geq \sigma_2 \geq \sigma_3$, this can be written as

$$\sigma_1 - \sigma_3 = C. \tag{5.6}$$

The constant, C, can be found by considering uniaxial tension. In a tension test, $\sigma_2 = \sigma_3 = 0$ and at yielding, $\sigma_1 = Y$ where Y is the yield strength. Substituting into Equation 5.6, $C = Y$. Therefore, the Tresca criterion may be expressed as

$$\sigma_1 - \sigma_3 = Y. \tag{5.7}$$

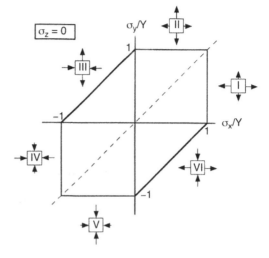

Figure 5.2. Plot of the yield locus for the Tresca criterion for $\sigma_z = 0$. The Tresca criterion predicts that the intermediate principal stress has no effect on yielding. For example, in sector I the value of σ_y has no effect on the value of σ_x required for yielding. Only if σ_y is negative or if it is greater than σ_x does it has an influence. In these cases, it is no longer the intermediate principal stress.

For pure shear, $\sigma_1 = -\sigma_3 = k$, where k is the shear yield strength. Substituting in Equation 5.7, $k = Y/2$ so

$$\sigma_1 - \sigma_3 = 2k = C. \tag{5.8}$$

EXAMPLE PROBLEM #5.1: Consider an isotropic material loaded so that the principal stresses coincide with the x, y, and z axes of the material. Assume that the Tresca yield criterion applies. Make a plot of the combinations of σ_y vs. σ_x that will cause yielding with $\sigma_z = 0$.

Solution: Divide the σ_y vs. σ_x stress space into six sectors, as shown in Figure 5.2. The following conditions are appropriate:

I $\sigma_x > \sigma_y > \sigma_z = 0$, so $\sigma_1 = \sigma_x$, $\sigma_3 = \sigma_z = 0$, so $\sigma_x = Y$

II $\sigma_y > \sigma_x > \sigma_z = 0$, so $\sigma_1 = \sigma_y$, $\sigma_3 = \sigma_z = 0$, so $\sigma_y = Y$

III $\sigma_y > \sigma_z = 0 > \sigma_x$, so $\sigma_1 = \sigma_y$, $\sigma_3 = \sigma_x$, so $\sigma_y - \sigma_x = Y$

VI $\sigma_x > \sigma_z = 0 > \sigma_y$, so $\sigma_1 = \sigma_x$, $\sigma_3 = \sigma_y$, so $\sigma_x - \sigma_y = Y$

These are plotted in Figure 5.2.

It seems reasonable to incorporate the effect of the intermediate principal stress into the yield criterion. One might try this by assuming that yielding depends on the average of the diameters of the three Mohr's circles, $[(\sigma_1 - \sigma_2) + (\sigma_2 - \sigma_3) + (\sigma_1 - \sigma_3)]/3$, but the intermediate stress term, σ_2, drops out of the average, $[(\sigma_1 - \sigma_2) + (\sigma_2 - \sigma_3) + (\sigma_1 - \sigma_3)]/3 = (2/3)(\sigma_1 - \sigma_3)$. Therefore, an average diameter criterion reduces to the Tresca criterion.

Von Mises Criterion

The effect of the intermediate principal stress can be included by assuming that yielding depends on the root-mean-square diameter of the three Mohr's circles.* This is the von Mises criterion, which can be expressed as

$$\{[(\sigma_2 - \sigma_3)^2 + (\sigma_3 - \sigma_1)^2 + (\sigma_1 - \sigma_2)^2]/3\}^{1/2} = C. \tag{5.9}$$

Note that each term is squared, so the convention $\sigma_1 \geq \sigma_2 \geq \sigma_3$ is not necessary. Again, the material constant, C, can be evaluated by considering a uniaxial tension test. At yielding, $\sigma_1 = Y$ and $\sigma_2 = \sigma_3 = 0$. Substituting, $[0^2 + (-Y)^2 + Y^2]/3 = C^2$, or $C = (2/3)^{1/3}Y$, so the equation is usually written as

$$(\sigma_2 - \sigma_3)^2 + (\sigma_3 - \sigma_1)^2 + (\sigma_1 - \sigma_2)^2 = 2Y^2. \tag{5.10}$$

For a state of pure shear, $\sigma_1 = -\sigma_3 = k$ and $\sigma_2 = 0$. Substituting into Equation 5.10, $(-k)^2 + [(-k) - k]^2 + k^2 = 2Y^2$, so

$$k = Y/\sqrt{3}. \tag{5.11}$$

Equation 5.10 can be simplified if one of the principal stresses is zero (plane-stress conditions). Substituting $\sigma_3 = 0$, $\sigma_1^2 + \sigma_2^2 - \sigma_1\sigma_2 = Y^2$, which is an ellipse. With further substitution of $a = \sigma_2/\sigma_1$,

$$\sigma_1 = Y/(1 - \alpha + \alpha^2)^{1/2}. \tag{5.12}$$

EXAMPLE PROBLEM #5.2: Consider an isotropic material loaded so that the principal stresses coincide with the x, y, and z axes. Assuming the von Mises yield criterion applies, make a plot of σ_y vs. σ_x yield locus with $\sigma_z = 0$.

Solution: Let $\sigma_y = \sigma_1$, $\sigma_y = \sigma_2$, and $\sigma_z = 0$. Now $\alpha = \sigma_2/\sigma_1$. Figure 5.3 results from substituting several values of a into Equation 6.12, solving for σ_x/Y and $\sigma_y/Y = \alpha\sigma_x/Y$, and then plotting.

EXAMPLE PROBLEM #5.3: For plane stress ($\sigma_3 = 0$), what is the largest possible ratio of σ_1/Y at yielding and at what stress ratio, α, does this occur?

Solution: Inspecting Equation 5.12, it is clear that maximum ratio of σ_1/Y corresponds to the minimum value $1 - \alpha + \alpha^2$. Differentiating and setting to zero, $d(1 - \alpha + \alpha^2)/d\alpha = -1 + 2\alpha$; $\alpha = -1/2$. Substituting into Equation 5.12,

$$\sigma_1/Y = [1 - 1/2 + (1/2)^2]^{-1/2} = \sqrt{(4/3)} = 1.155.$$

---

* This is equivalent to assuming that yielding occurs when the elastic distortional strain energy reaches a critical value.

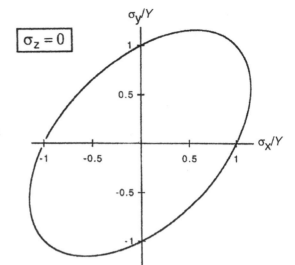

Figure 5.3. The von Mises criterion with $\sigma_z = 0$ plots as an ellipse.

The von Mises yield criterion can also be expressed in terms of stresses that are not principal stresses. In this case, it is necessary to include shear terms,

$$(\sigma_y - \sigma_z)^2 + (\sigma_z - \sigma_x)^2 + (\sigma_x - \sigma_y)^2 + 6\left(\tau_{yz}^2 + \tau_{zx}^2 + \tau_{xy}^2\right) = 2Y^2, \qquad (5.13)$$

Flow Rules

When a material yields, the ratio of the resulting strains depends on the stress state that causes yielding. The general relations between plastic strains and the stress states are called the *flow rules*. They may be expressed as

$$d\varepsilon_{ij} = d\lambda(\partial f/\partial\sigma_{ij}), \qquad (5.14)$$

where f is the yield function corresponding to the yield criterion of concern and $d\lambda$ is a constant that depends on the shape of the stress strain curve.* For the von Mises criterion, we can write $f = [(\sigma_2 - \sigma_3)^2 + (\sigma_3 - \sigma_1)^2 + (\sigma_1 - \sigma_2)^2]/4$. In this case, Equation 5.14 results in

$$d\varepsilon_1 = d\lambda[2(\sigma_1 - \sigma_2) - 2(\sigma_1 - \sigma_1)]/4 = d\lambda\cdot[\sigma_1 - (\sigma_2 + \sigma_3)/2]$$
$$d\varepsilon_2 = d\lambda[\sigma_2 - (\sigma_3 + \sigma_1)/2] \qquad (5.15)$$
$$d\varepsilon_3 = d\lambda[\varepsilon_3 - (\sigma_1 + \sigma_2)/2].^{**}$$

---

* The constant, $d\lambda$, can be expressed as $d\lambda = d\bar{\varepsilon}/d\bar{\sigma}$, which is the inverse slope of the effective stress-strain curve at the point where the strains are being evaluated.

** Equations 5.15 parallel Hooke's law equations (Equation 2.2) where $d\lambda = d\bar{\varepsilon}/d\bar{\sigma}$ replaces $1/E$ and $1/2$ replaces Poisson's ratio, υ. For this reason, it is sometimes said that the "plastic Poisson's ratio" is $1/2$.

These are known as the Levy-Mises equations. Even though $d\lambda$ is not usually known, these equations are useful for finding the ratio of strains that result from a known stress state or the ratio of stresses that correspond to a known strain state.

> **EXAMPLE PROBLEM #5.4:** Find the ratio of the principal strains that result from yielding if the principal stresses are $\sigma_y = \sigma_x/4$, $\sigma_z = 0$. Assume the von Mises criterion.
>
> *Solution:* (a) According to Equation 5.15, $d\varepsilon_1 : d\varepsilon_2 : d\varepsilon_3 = [\sigma_1 - (\sigma_2 + \sigma_3)/2] : [\sigma_2 - (\sigma_3 + \sigma_1)/2] : [\sigma_3 - (\sigma_1 + \sigma_2)/2] = (7/8)\sigma_1 : (-1/4)\sigma_1 : (-5/8)\sigma_1 = 7 : -2 : -5$.

The flow rules for the Tresca criterion can be found by applying Equation 5.14 with $f = \sigma_1 - \sigma_3$. Then $d\varepsilon_1 = d\lambda$, $d\varepsilon_2 = 0$, and $d\varepsilon_3 = -d\lambda$, or

$$d\varepsilon_1 : d\varepsilon_2 : d\varepsilon_3 = 1 : 0 : -1. \tag{5.16}$$

> **EXAMPLE PROBLEM #5.5:** Find the ratio of the principal strains that result from yielding if the principal stresses are $\sigma_y = \sigma_x/4$, $\sigma_z = 0$, assuming the Tresca criterion.
>
> *Solution:* Here $d\varepsilon_1 = d\varepsilon_x$, $d\varepsilon_2 = d\varepsilon_y$, and $d\varepsilon_3 = d\varepsilon_z$, so $d\varepsilon_x : d\varepsilon_y : d\varepsilon_z = 1 : 0 : -1$.

> **EXAMPLE PROBLEM #5.6:** Circles were printed on the surface of a part before it was deformed. Examination after deformation revealed that the principal strains in the sheet are $\varepsilon_1 = 0.18$ and $\varepsilon_2 = 0.078$. Assume that the tools did not touch the surface of concern and that the ratio of stresses remained constant during the deformation. Using the von Mises criterion, find the ratio of the principal stresses.
>
> *Solution:* First, realize that with a constant ratio of stresses, $d\varepsilon_2 : d\varepsilon_1 = \varepsilon_2 : \varepsilon_1$. Also, if the tools did not touch the surface, it can be assumed that $\sigma_3 = 0$. Then according to Equation 5.15, $\varepsilon_2/\varepsilon_1 = d\varepsilon_2/d\varepsilon_1 = (\sigma_2 - \sigma_1/2)/(\sigma_1 - \sigma_2/2)$. Substituting, $\alpha = \sigma_2/\sigma_1$, and $\rho = \varepsilon_2/\varepsilon_1$, $(\alpha - 1/2)/(1 - \alpha/2) = \rho$. Solving for α,

$$\alpha = (\rho + 1/2)/(\rho/2 + 1). \tag{5.17}$$

> Now substituting $\rho = 0.078/0.18 = 0.4333$, $\alpha = 0.7671$.

> **EXAMPLE PROBLEM #5.7:** Draw the three-dimensional Mohr's circle diagrams for the stresses and plastic strains in a body loaded under plane-strain tension, $\varepsilon_y = 0$, with $\sigma_z = 0$. Assume the von Mises yield criterion.

Figure 5.4. Mohr's strain and stress circles for plane strain.

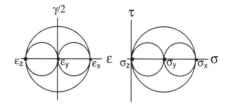

Solution: From the flow rules, $\varepsilon_y = 0 = \lambda[\sigma_y - (1/2)(\sigma_x + \sigma_z)]$ with $e_y = 0$ and $\sigma_z = 0$, $\sigma_y = 1/2(\sigma_x)$. The Mohr's stress circles are determined by the principal stresses, σ_x, $\sigma_y = 1/2(\sigma_x)$ and $\sigma_z = 0$. For plastic flow, $\varepsilon_x + \varepsilon_y + \varepsilon_z = 0$, so with $\varepsilon_y = 0$, $\varepsilon_z = -\varepsilon_x$. The Mohr's strain circles (Figure 5.4) are determined by the principal strains, ε_x, $\varepsilon_y = 0$, and $\varepsilon_z = -\varepsilon_x$.

Principle of Normality

The flow rules may be represented by the *principle of normality*. According to this principle, if a normal is constructed to the yield locus at the point of yielding, the strains that result from yielding are in the same ratio as the stress components of the normal. This is illustrated in Figure 5.5. A corollary is that for a σ_1 vs. σ_2 yield locus with $\sigma_3 = 0$,

$$d\varepsilon_1/d\varepsilon_2 = -\partial\sigma_2/\partial\sigma_1, \qquad (5.18)$$

where $\partial\sigma_2/\partial\sigma_1$ is the slope of the yield locus at the point of yielding. It should be noted that Equations 5.14 and 5.17 are general and can be used with other yield criteria, including ones formulated to account for anisotropy and for pressure-dependent yielding.

Figure 5.6 shows how different shape changes result from different loading paths. The components of the normal at A are $d\sigma_y/d\sigma_x = 1$, so $\varepsilon_y = \varepsilon_x$. At B, the normal has a slope of $d\sigma_y = 0$, so $\varepsilon_y = 0$. For uniaxial tension at C, the slope of the normal is $d\sigma_y/d\sigma_x = -1/2$ which corresponds to $\varepsilon_y = (-1/2)\varepsilon_x$.

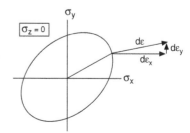

Figure 5.5. The ratios of the strains resulting from yielding are in the same proportion as the components of a vector normal to the yield surface at the point of yielding.

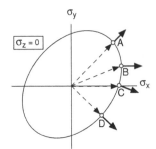

Figure 5.6. The ratio of strains resulting from yielding along several loading paths. At A, $\varepsilon_y = \varepsilon_x$; at B, $\varepsilon_y = 0$; at C (uniaxial tension) $\varepsilon_y = -1/2e_x$; and at D, $\varepsilon_y = -\varepsilon_x$.

At D, $d\sigma_y/d\sigma_x = -1$, so $\varepsilon_y = -\varepsilon_x$. Figure 5.7 is a representation of the normality principle applied to the Tresca criterion. All stress states along a straight edge cause the same ratio of plastic strains. The shape changes corresponding to the corners are ambiguous because it is ambiguous which stress component is σ_{max} and which stress component is σ_{min}. For example, with yielding under biaxial tension, $\sigma_x = \sigma_y$, $0 \le \varepsilon_y/\varepsilon_x \le \infty$.

Effective Stress and Effective Strain

The concepts of effective stress and effective strain are necessary for analyzing the strain hardening that occurs on loading paths other than uniaxial tension. Effective stress, $\overline{\sigma}$, and effective strain, $\overline{\varepsilon}$, are defined so that:

1) $\overline{\sigma}$ and $\overline{\varepsilon}$ reduce to σ_x and ε_x in an x-direction tension test.
2) The incremental work per volume done in deforming a material plastically is $dw = \overline{\sigma}\, d\overline{\varepsilon}$.
3) Furthermore it is usually assumed that the $\overline{\sigma}$ vs. $\overline{\varepsilon}$ curve describes the strain hardening for loading under a constant stress ratio, α, regardless of α.*

These orientation changes (texture development) depend on the loading path. However, the dependence of strain hardening on the loading path is significant only at large strains.

The effective stress, $\overline{\sigma}$, is the function of the applied stresses that determine whether yielding occurs. When $\overline{\sigma}$ reaches the current flow stress, plastic deformation will occur. For the von Mises criterion,

$$\overline{\sigma} = (1/\sqrt{2})[(\sigma_2 - \sigma_3)^2 + (\sigma_3 - \sigma_1)^2 + (\sigma_1 - \sigma_2)^2]^{1/2}, \tag{5.19}$$

and for the Tresca criterion,

$$\overline{\sigma} = (\sigma_1 - \sigma_3). \tag{5.20}$$

---

* It should be understood that for large strains, the $\overline{\sigma}$ vs. $\overline{\varepsilon}$ curves do depend on the loading path because of orientation changes within the material.

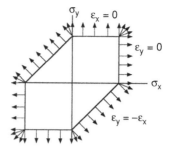

Figure 5.7. The normality principle applied to the Tresca yield criterion. All stress states on the same side of the locus cause the same shape change. The shape changes at the corners are ambiguous. Biaxial tension can produce any shape change from $\varepsilon_y = 0$ to $\varepsilon_x = 0$.

Note that in a tension test, the effective stress reduces to the tensile stress so both criteria predict yielding when $\bar{\sigma}$ equals the current flow stress.

The effective strain, $\bar{\varepsilon}$, is a mathematical function of the strain components, defined in such a way that $\bar{\varepsilon}$ reduces to the tensile strain in a tension test and that the plastic work per volume is

$$dw = \bar{\sigma}d\bar{\varepsilon} = \sigma_1 d\varepsilon_1 + \sigma_2 d\varepsilon_2 + \sigma_3 d\varepsilon_3. \tag{5.21}$$

For the von Mises criterion, $d\bar{\varepsilon}$, can be expressed as

$$d\bar{\varepsilon} = (\sqrt{2}/3)[(d\varepsilon_2 - d\varepsilon_3)^2 + (d\varepsilon_3 - d\varepsilon_1)^2 + (d\varepsilon_1 - d\varepsilon_2)^2]^{1/2}, \tag{5.22}$$

or as

$$d\bar{\varepsilon} = (2/3)^{1/3}\left(d\varepsilon_1^2 + d\varepsilon_2^2 + d\varepsilon_3^2\right)^{1/2}, \tag{5.23}$$

which is completely equivalent.

EXAMPLE PROBLEM #5.8: Show that $d\bar{\varepsilon}$ in Equations 5.22 and 5.23 reduce to $d\varepsilon_1$ in a one-direction tension test.

Solution: In a one-direction tension test, $d\varepsilon_2 = d\varepsilon_3 = -(1/2)d\varepsilon_1$. Substituting into Equation 5.22, $d\bar{\varepsilon} = (\sqrt{2}/3)\{(0)^2 + [-(1/2)d\varepsilon_1 - d\varepsilon_1]^2 + [d\varepsilon_1 - (-1/2)d\varepsilon_1]^2\}^{1/2}, = (\sqrt{2}/3)[(9/4)d\varepsilon_1^2 + (9/4)d\varepsilon_1^2]^{1/2} = d\varepsilon_1$ Substituting $d\varepsilon_2 = d\varepsilon_3 = -(1/2)d\varepsilon_1$ into Equation 5.22, $d\bar{\varepsilon} = (2/3)[d\varepsilon_1^2 + (-1/2d\varepsilon_1)^2 + (-1/2d\varepsilon_1)^2]^{1/2} = \sqrt{(2/3)}[d\varepsilon_1^2 + d\varepsilon_1^2/4 + d\varepsilon_1^2/4]^{1/2}d = \varepsilon_1$.

If the straining is proportional (constant ratios of $d\varepsilon_1 : d\varepsilon_2 : d\varepsilon_3$), the total effective strain can be expressed as

$$\bar{\varepsilon} = \left[(2/3)\left(\varepsilon_1^2 + \varepsilon_2^2 + \varepsilon_3^2\right)\right]^{1/2}. \tag{5.24}$$

If the straining is not proportional, $\bar{\varepsilon}$ must be found by integrating $d\bar{\varepsilon}$ along the strain path.

The effective strain (and stress) may also be expressed in terms of non-principal strains (and stresses). For von Mises,

$$\bar{\varepsilon} = \left[(2/3)\left(e_x^2 + \varepsilon_y^2 + \varepsilon_z^2\right) + (1/3)\left(\gamma_{yz}^2 + \gamma_{zx}^2 + \gamma_{xy}^2\right) \right]^{1/2}, \tag{5.25}$$

and

$$\bar{\sigma} = (1/\sqrt{2})\left[(\sigma_y - \sigma_z)^2 + \left(\sigma_z - \sigma_x^2 + (\sigma_x - \sigma_y)^2 + 6\left(\tau_{yz}^2 + \tau_{zx}^2 + \tau_{xy}^2\right) \right]^{1/2}. \tag{5.26}$$

For the Tresca criterion, the effective strain is the absolutely largest principal strain,

$$d\bar{\varepsilon} = |d\varepsilon_i|_{\max}. \tag{5.27}$$

Although the Tresca effective strain is not widely used, it is of value because it is so extremely simple to find. Furthermore, it is worth noting that the value of the von Mises effective strain can never differ from it by more than 15%. Always,

$$|d\varepsilon_i|_{\max} \le d\bar{\varepsilon}_{\text{mises}} \le 1.15|d\varepsilon_i|_{\max}. \tag{5.28}$$

Equation 5.28 provides a simple check when calculating the von Mises effective strain. If one calculates a value of $d\bar{\varepsilon}$ for von Mises that does not fall within the limits of Equation 5.28, a mistake has been made.

As a material deforms plastically, the level of stress necessary to continue deformation increases. It is postulated that the strain hardening depends only on $\bar{\varepsilon}$. In that case, there is a unique relation,

$$\bar{\sigma} = f(\bar{\varepsilon}). \tag{5.29}$$

Because for a tension test $\bar{\varepsilon}$ is the tensile strain and $\bar{\sigma}$ is the tensile stress, the $\sigma - \varepsilon$ curve in a tension test is the $\bar{\sigma} - \bar{\varepsilon}$ curve. Therefore, the stress-strain curve in a tension test can be used to predict the stress-strain behavior under other forms of loading.

> **EXAMPLE PROBLEM #5.9:** The strains measured on the surface of a piece of sheet metal after deformation are $\varepsilon_1 = 0.182$ and $\varepsilon_2 = -0.035$. The stress-strain curve in tension can be approximated by $\sigma = 30 + 40\varepsilon$. Assume the von Mises criterion and assume that the loading was such that the ratio of $\varepsilon_2/\varepsilon_1$ was constant. Calculate the levels of σ_1 and σ_2 reached before unloading.
>
> *Solution:* First, find the effective strain. With constant volume, $\varepsilon_3 = -\varepsilon_1 - \varepsilon_2 = -0.182 + 0.035 = -0.147$. The von Mises effective strain is $\bar{\varepsilon} = [(2/3)(\varepsilon_1^2 + \varepsilon_2^2 + \varepsilon_3^2)]^{1/2} = [(2/3)(0.182^2 + 0.035^2 + 0.147^2)]^{1/2} = 0.193$. (Note that this is larger than the largest principal strain, 0.182, and smaller than 1.15×0.182). Because the tensile stress-strain curve is the effective

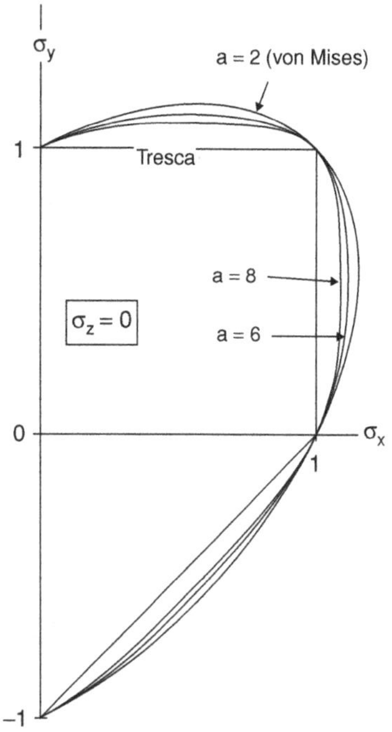

Figure 5.8. Yield loci for $(\sigma_2 - \sigma_3)^a + (\sigma_3 - \sigma_1)^a + (\sigma_1 - \sigma_2)^a = 2Y^a$ with several values of a. Note that the von Mises criterion corresponds to $a = 2$ and the Tresca criterion to $a = 1$.

stress-effective strain relation, $\bar{\sigma} = 30 + 40\bar{\varepsilon} = 30 + 40 \times 0.193 = 37.7$. At the surface $\sigma_z = 0$, so the effective stress function can be written as $\bar{\sigma}/\sigma_1 = (1/\sqrt{2})[\alpha^2 + 1 + (1 - \alpha)^2]^{1/2} = (1 - \alpha + \alpha^2)^{1/2}$. From the flow rules with $\sigma_z = 0$, $\rho = d\varepsilon_2/d\varepsilon_1 = (\alpha - 1/2)/(1 - \alpha/2)$. Solving for α, $\bar{\varepsilon} = (\rho + 1/2)/(1 + \rho/2)$. Substituting $\rho = -0.035/0.182 = -0.192$, $\alpha = (-0.192 + 1/2)/(1 - 0.192/2) = +0.341$, $\bar{\sigma}/\sigma_1 = [1 - \alpha + \alpha^2]1/2 = [1 - 0.341 + 0.341^2]^{1/2} = 0.881$. $\sigma_1 = 0.881/\bar{\sigma} = 37.7/0.881 = 42.8$ and $\sigma_2 = \alpha\sigma_1 = 0.341 \times 42.8 = 14.6$

Other Isotropic Yield Criteria

von Mises and Tresca are not the only possible isotropic criteria. Both experimental data and theoretical analysis based on a crystallographic model tend to lie between the two and can be represented by[*]

$$|\sigma_2 - \sigma_3|^a + |\sigma_3 - \sigma_1|^a + |\sigma_1 - \sigma_2|^a = 2Y^a. \tag{5.30}$$

This criterion reduces to von Mises for $a = 2$ and $a = 4$, and to Tresca for $a = 1$ and $a \to \infty$. For values of the exponent greater than 4, this criterion predicts yield loci between Tresca and von Mises, as shown in Figure 5.8. If the exponent, a, is an even integer, this criterion can be written without the absolute magnitude signs as

$$(\sigma_2 - \sigma_3)^a + (\sigma_3 - \sigma_1)^a + (\sigma_1 - \sigma_2)^a = 2Y^a. \tag{5.31}$$

* W. F. Hosford, *J. Appl. Mech.* (*Trans. ASME ser E.*) v. 39E (1972).

Theoretical calculations based on {111} <110> slip suggest an exponent of $a = 8$ for fcc metals. Similar calculations suggest that $a = 6$ for bcc. These values fit experimental data as well.

EXAMPLE PROBLEM #5.10: Derive the flow rules for the high exponent yield criterion, Equation, 5.29.

Solution: Expressing the yield function as $f = (\sigma_y - \sigma_z)^a + (\sigma_z - \sigma_x)^a + (\sigma_x - \sigma_y)^a$, the flow rules can be found from $d\varepsilon_{ij} = d\lambda(df/d\sigma_{ij})$. $d\varepsilon_x = d\lambda[a(\sigma_x - \sigma_z)^{a-1} + a(\sigma_x - \sigma_y)^{a-1}]$, $d\varepsilon_y = d\lambda[a(\sigma_y - \sigma_z)^{a-1} + a(\sigma_y - \sigma_z)^{a-1}]$, and $d\varepsilon_z = d\lambda[a(\sigma_z - \sigma_x)^{a-1} + a(\sigma_z - \sigma_y)^{a-1}]$. These can be expressed as

$$d\varepsilon_x : d\varepsilon_y : d\varepsilon_z = [(\sigma_x - \sigma_z)^{a-1} + (\sigma_x - \sigma_y)^{a-1}] :$$
$$[(\sigma_y - \sigma_x)^{a-1} + (\sigma_y - \sigma_z)^{a-1}] : [(\sigma_z - \sigma_x)^{a-1} + (\sigma_z - \sigma_y)^{a-1}]$$

$$(5.32)$$

Effect of Strain Hardening on the Yield Locus

According to the *isotropic hardening* model, the effect of strain hardening is simply to expand the yield locus without changing its shape. The stresses for yielding are increased by the same factor along all loading paths. This is the basic assumption that $\bar{\sigma} = f(\bar{\varepsilon})$. The isotropic hardening model can be applied to anisotropic materials. It does not imply that the material is isotropic.

An alternative model is *kinematic hardening*. According to this model, plastic deformation simply shifts the yield locus in the direction of the loading path without changing its shape or size. If the shift is large enough, unloading may actually cause plastic deformation. The kinematic model is probably better for describing small strains after a change in load path. However, the isotropic model is better for describing behavior during large strains after a change of strain path. Figure 5.9 illustrates both models.

Notes

Otto Z. Mohr (1835–1918) worked as a civil engineer, designing bridges. At 32, he was appointed a professor of engineering mechanics at Stuttgart Polytecknium. Among other contributions, he here devised the graphical method of analyzing the stress at a point.

He then extended Coulomb's idea that failure is caused by shear stresses into a failure criterion based on maximum shear stress, or diameter of the largest circle.* For cast iron, he proposed the different fracture stresses in tension, shear, and compression could be combined into a single diagram in which the tangents form an envelope of safe stress combinations. This is essentially the Tresca yield criterion. It may be noted that early workers used

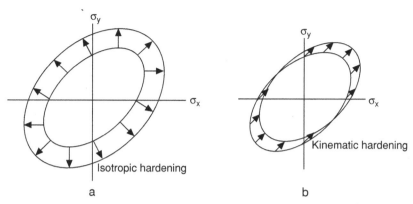

Figure 5.9. The effect of strain hardening on the yield locus. The isotropic model (a) predicts an expansion of the locus. The kinematic hardening model predicts a translation of the locus in the direction of the loading path.

the term "failure criteria," which failed to distinguish between fracture and yielding.

Lamé assumed a maximum stress theory of failure. However, later a maximum strain theory of which Poncelet and Saint-Venant were proponents became generally accepted. They proposed that failure would occur when a critical strain was reached, regardless of the stress state.

In letters to William Thompson, John Clerk Maxwell (1831–1879) proposed that "strain energy of distortion" was critical but he never published this idea, and it was forgotten. In 1904, M. T. Huber first formulated the expression for "distortional strain energy,"

$$U = [1/(12G)][(\sigma_2 - \sigma_3)^2 + (\sigma_3 - \sigma_1)^2 + (\sigma_1 - \sigma_2)^2] \text{ where } U = \sigma_{yp}^2/(6G).$$

The same idea was independently developed by von Mises (*Göttinger. Nachr.* p. 582,1913), for whom the criterion is generally called. It is also referred to by the names of several people who independently proposed it: Huber, Hencky, as well as Maxwell. It is also known as the "maximum distortional energy" theory and the "octahedral shear stress" theory. The first name reflects that the elastic energy in an isotropic material associated with shear (in contrast to dilatation) is proportional to the left side of Equation 5.10. The second name reflects that the shear terms, $(\sigma_2 - \sigma_3)$, $(\sigma_3 - \sigma_1)$, and $(\sigma_1 - \sigma_2)$, can be represented as the edges of an octahedron in principal stress space.

In 1868, Tresca presented two notes to the French Academy. From these, Saint-Venant established the first theory of plasticity based on the assumptions that:

1) Plastic deformation does not change the volume of a material,
2) Directions of principal stresses and principal strains coincide,
3) The maximum shear stress at a point is a constant.

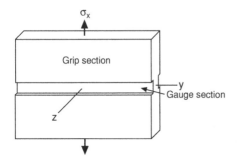

Figure 5.10. Plane-strain tensile specimen. Lateral contraction of material in the groove is constrained by material outside the groove.

The Tresca criterion is also called the *Guest* or the *"maximum shear stress"* criterion.

REFERENCES

W. F. Hosford and R. M. Caddell, *Metal Forming: Mechanics and Metallurgy, 3rd Ed.*, Cambridge U. Press, 2007.
R. Hill, *The Mathematical Theory of Plasticity*, Oxford (1950).
W. A. Backofen, *Deformation Processing*, Addison-Wesley (1972).
W. F. Hosford, *Mechanical Behavior of Materials*, Cambridge U. Press, (2005).

Problems

1. For the von Mises yield criterion with $\sigma_z = 0$, calculate the values of σ_x/Y at yielding if

   a. $\alpha = 1/2$ b. $\alpha = 1$ c. $\alpha = -1$ d. $\alpha = 0$ where $\alpha = \sigma_y/\sigma_x$.

2. For each of the values of α in Problem 1, calculate the ratio $\rho = d\varepsilon_y d\varepsilon_x$.

3. Repeat Problems 1 and 2 assuming the Tresca criterion instead of the von Mises criterion.

4. Repeat Problems 1 and 2 assuming the following yield criterion:

   $$(\sigma_2 - \sigma_3)^a + (\sigma_3 - \sigma_1)^a + (\sigma_1 - \sigma_2)^a = 2Y^a \quad \text{where } a = 8.$$

5. Consider a plane-strain tension test (Figure 5.10) in which the tensile stress is applied along the x-direction. The strain, ε_y, is zero along the transverse direction, and the stress in the z-direction vanishes. Assuming the von Mises criterion, write expressions for:

   A. $\bar{\sigma}$ as a function of σ_x, and

   B. $d\bar{\varepsilon}$ as a function of $d\varepsilon_x$.

   C. Using the results of parts A and B, write an expression for the incremental work per volume, dw, in terms of σ_x and $d\varepsilon_x$.

   D. Derive an expression for σ_x as a function of ε_x in such a test, assuming that the strain hardening can be expressed by $\bar{\sigma} = K\bar{\varepsilon}^n$.

6. A 1.00-cm diameter circle was printed onto the surface of a sheet of steel before forming. After forming, the circle was found to be an ellipse with major and minor diameters of 1.18 cm and 1.03 cm, respectively. Assume

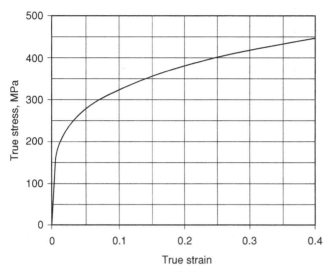

Figure 5.11. True tensile stress-strain curve for the steel in Problem 6.

that both sets of measurements were made when the sheet was unloaded, that during forming the stress perpendicular to the sheet surface was zero, and that the ratio, α, of the stresses in the plane of the sheet remained constant during forming. The tensile stress-strain curve for this steel is shown in Figure 5.11. Assume the von Mises yield criterion.

A. What were the principal strains, ε_1, ε_2, and ε_3?

B. What was the effective strain, $\bar{\varepsilon}$?

C. What was the effective stress, $\bar{\sigma}$?

D. Calculate the ratio, $\rho = \varepsilon_2/\varepsilon_1$ and use this to find the ratio, $\alpha = \sigma_2/\sigma_1$. (Take σ_1 and σ_2 respectively as the larger and smaller of the principal stresses in the plane of the sheet.)

E. What was the level of σ_1?

7. Measurements on the surface of a deformed sheet after unloading indicate that $e_1 = 0.154$ and $e_2 = 0.070$. Assume that the von Mises criterion is appropriate and that the loading was proportional (i.e., the ratio $\alpha = \sigma_y/\sigma_x$ remained constant during loading). It has been found that the tensile stress–strain relationship for this alloy can be approximated by $\sigma = 150 + 185\varepsilon$ (Figure 5.12), where σ is the true stress in MPa and ε is the true strain.

Figure 5.12. True tensile stress-strain curve for the steel in Problem 7.

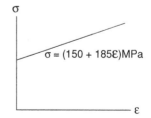

    A. What was the effective strain?

    B. What was the effective stress?

    C. What was the value of the largest principal stress?

8. The following yield criterion has been proposed for an isotropic material: "Yielding will occur when the sum of the diameters of the largest and second largest Mohr's circles reaches a critical value. Defining $\sigma_1 \geq \sigma_2 \geq \sigma_3$, this can be expressed mathematically as: If $(\sigma_1 - \sigma_2) \geq (\sigma_2 - \sigma_3)$, $(\sigma_1 - \sigma_2) + (\sigma_1 - \sigma_3) = C$ or $2\sigma_1 - \sigma_2 - \sigma_3 = C$, but if $(\sigma_2 - \sigma_3) \geq (\sigma_1 - \sigma_2)$, $(\sigma_1 - \sigma_3) + (\sigma_2 - \sigma_3) = C$ or $\sigma_1 + \sigma_2 - 2\sigma_3 = C$.

    A. Evaluate C in terms of the yield strength, Y, in uniaxial tension or the yield strength, $-Y$, in compression

    Plot the yield locus as σ_x vs. σ_y for $\sigma_z = 0$ where σ_x, σ_y, and σ_z are principal stresses. [Hint: For each region in σ_x vs. σ_y stress space, determine whether Equations applies.]

9. Consider a long thin-walled tube, capped at both ends. It is made from a steel with a yield strength of 40,000 psi. Its length is 60 in., its diameter is 2.0 in. and the wall thickness is 0.015 in. The tube is loaded under an internal pressure, P, and a torque of 1500 in. lbs is applied.

    A. What internal pressure can it withstand without yielding according to Tresca?

    B. What internal pressure can it withstand without yielding according to von Mises?

10. In the flat rolling of a sheet or plate, the width does not appreciably change. A sheet of aluminum is rolled from 0.050 in. to 0.025 in. thickness. Assume the von Mises criterion.

    A. What is the effective strain, $\bar{\varepsilon}$, caused by the rolling?

    B. What strain in a tension test (if it were possible) would cause the same amount of strain hardening?

11. A piece of ontarium (which has a tensile yield strength of $Y = 700$ MPa) was loaded in such away that the principal stresses, σ_x, σ_y, and σ_z, were in the ratio of 1:0:−0.25. The stresses were increased until plastic deformation occurred.

    A. Predict the ratio of the principal strains, $\rho = \varepsilon_y/\varepsilon_x$, resulting from yielding according to von Mises.

    B. Predict the value of $\rho = \varepsilon_y/\varepsilon_x$ esulting from yielding according to Tresca.

    C. Predict the value of σ_x when yielding occurred according to von Mises.

    D. Predict the value of σ_x when yielding occurred according to Tresca.

12. A new yield criterion has been proposed for isotropic materials. It states that yielding will occur when the diameter of Mohr's largest circle plus half of the diameter of Mohr's second largest circle equals a critical value.

This criterion can be expressed mathematically, following the convention that $\sigma_1 \geq \sigma_2 \geq \sigma_3$, as

$$(\sigma_1 - \sigma_3) + 1/2(\sigma_1 - \sigma_2) = C \quad \text{if } (\sigma_1 - \sigma_2) \geq (\sigma_2 - \sigma_3)$$

and

$$(\sigma_1 - \sigma_3) + 1/2(\sigma_2 - \sigma_3) = C \quad \text{if } (\sigma_2 - \sigma_3) \geq (\sigma_1 - \sigma_2).$$

A. Evaluate C in terms of the tensile (or compressive) yield strength, Y.

B. Let x, y, and z be directions of principal stress, and let $\sigma_z = 0$. Plot the σ_y vs. σ_x yield locus, that is, plot the values of σ_y/Y and σ_x/Y that will lead to yielding according to this criterion.

[Hint: Consider different loading paths (ratios of σ_y/σ_x), and for each decide which stress (σ_1, σ_2, or σ_3) corresponds to (σ_x, σ_y or $\sigma_z = 0$), then determine whether $(\sigma_1 - \sigma_2) \geq (\sigma_2 - \sigma_3)$, substitute s_X, s_y, and 0 into the appropriate expression, solve and, finally, plot.]

13. The tensile yield strength of an aluminum alloy is 14,500 psi. A sheet of this alloy is loaded under plane-stress conditions ($\sigma_3 = 0$) until it yields. On unloading, it is observed that $\varepsilon_1 = 2\varepsilon_2$ and both ε_1 and ε_2 are positive.

A. Assuming the von Mises yield criterion, determine the values of σ_1 and σ_2 at yielding.

B. Sketch the yield locus and show where the stress state is located on the locus.

14. In a tension test of an anisotropic sheet, the ratio of the width strain to the thickness strain, $\varepsilon_w/\varepsilon_t$, is R.

A. Express the ratio $\varepsilon_2/\varepsilon_1$ of the strains in the plane of the sheet in terms of R. Take the 1-direction as the rolling direction, the 2-direction as the width direction in the tension test, and the 3-direction as the thickness direction.

B. There is a direction, x, in the plane of the sheet along which $\varepsilon_x = 0$. Find the angle, θ, between x and the tensile axis.

15. Redo Problem 6 assuming the Tresca criterion instead of the, von Mises criterion.

16. The total volume of a foamed material decreases when it plastically deforms in tension.

A. What does this imply about the effect of $\sigma_H = \sigma_1 + \sigma_2 + \sigma_3)/3$ on the shape of the yield surface in σ_1, σ_2, σ_3 space?

B. Would the absolute magnitude of the yield stress in compression be greater, smaller, or the same as the yield strength in tension?

C. When it yields in compression, would the volume increase, decrease, or remain constant?

6 Strain-Rate and Temperature Dependence of Flow Stress

Introduction

An increase of strain rate raises the flow stress of most materials. The amount of the effect depends on the material and the temperature. In most metallic materials, the effect near room temperature is small and is often neglected. A factor of ten increase in strain rate may raise the level of the stress-strain curve by only 1% or 2%. On the other hand, at elevated temperatures the effect of strain rate on flow stress is much greater. Increasing the strain rate by a factor of ten may raise the stress-strain curve by 50% or more.

Strain localization occurs very slowly in materials that have a high strain-rate dependence because less-strained regions continue to deform. Under certain conditions, the rate dependence is large enough for materials to behave *superplastically.* Tensile elongations of 1000% are possible.

There is a close coupling of the effects of temperature and strain rate on the flow stress. Increased temperatures have the same effects as deceased temperatures. This coupling can be understood in terms of the Arrhenius rate equation.

Strain-Rate Dependence of Flow Stress

The average strain rate during most tensile tests is in the range of 10^{-3} to 10^{-2}/s. If it takes 5 min during the tensile test to reach a strain of 0.3, the average strain rate is $\dot{\varepsilon} = 0.3/(5 \times 60) = 10^{-3}$/s. At a strain rate of $\dot{\varepsilon} = 10^{-2}$/s a strain of 0.3 will occur in 30 seconds. For many materials, the effect of the strain rate on the flow stress, σ, at a fixed strain and temperature can be described by a power-law expression,

$$\sigma = C\dot{\varepsilon}^m, \tag{6.1}$$

where the exponent, m, is called the *strain-rate sensitivity.* The relative levels of stress at two strain rates (measured at the same total strain) is given by

$$\sigma_2/\sigma_1 = (\dot{\varepsilon}_2/\dot{\varepsilon}_1)^m, \tag{6.2}$$

or $\ln(\sigma_2/\sigma_1) = m\ln(\dot{\varepsilon}_2/\dot{\varepsilon}_1)$. If σ_2 is not much greater than σ_1,

$$\ln(\sigma_2/\sigma_1) \approx \Delta\sigma/\sigma. \tag{6.3}$$

Equation 6.2 can be simplified to

$$\Delta\sigma/\sigma \approx m\ln(\dot{\varepsilon}_2/\dot{\varepsilon}_1) = 2.3m\log(\dot{\varepsilon}_2/\dot{\varepsilon}_1). \tag{6.4}$$

At room temperature, the values of m for most engineering metals are between -0.005 and $+0.015$, as shown in Table 6.1.

Consider the effect of a ten-fold increase in strain rate, $(\dot{\varepsilon}_2/\dot{\varepsilon}_1 = 10)$ with $m = 0.01$. Equation 6.4 predicts that the level of the stress increases by only $\Delta\sigma/\sigma = 2.3(0.01)(1) = 2.3\%$. This increase is typical of room-temperature tensile testing. It is so small that the effect of strain rate is often ignored. A plot of Equation 6.2 in Figure 6.1 shows how the relative flow stress depends on the strain rate for several levels of m. The increase of flow stress, $\Delta\sigma/\sigma$, is small unless either m or $(\dot{\varepsilon}_2/\dot{\varepsilon}_1)$ is large.

> **EXAMPLE PROBLEM #6.1:** The strain-rate dependence of a zinc alloy can be represented by Equation 6.1 with $m = 0.07$. What is the ratio of the flow stress at $\varepsilon = 0.10$ for a strain rate of 10^3/s to that at $\varepsilon = 0.10$ for a strain rate of 10^{-3}/s? Repeat for a low-carbon steel with $m = 0.01$.
>
> *Solution:* For zinc $(m = 0.07)$, $\sigma_2/\sigma_1 = (C\dot{\varepsilon}_2^m)/(C\dot{\varepsilon}^m) = (\dot{\varepsilon}_2/\dot{\varepsilon}_1)m = (10^3/10^{-3})^{0.07} = (10^6)^{0.07} = 2.63$. For steel $(m = 0.01)$, $\sigma_2/\sigma_1 = (10^6)^{0.01} = 1.15$.

> **EXAMPLE PROBLEM #6.2:** The tensile stress in one region of an HSLA steel sheet $(m = 0.005)$ is 1% greater than in another region. What is the ratio of the strain rates in the two regions? Neglect strain hardening. What would be the ratio of the strain rates in the two regions for a titanium alloy $(m = 0.02)$?

Table 6.1. *Typical values of the strain-rate exponent, m, at room temperature*

Material	m
low-carbon steels	0.010 to 0.015
HSLA steels	0.005 to 0.010
austenitic stainless steels	-0.005 to $+0.005$
ferritic stainless steels	0.010 to 0.015
copper	0.005
70/30 brass	-0.005 to 0
aluminum alloys	-0.005 to $+0.005$
α-titanium alloys	0.01 to 0.02
zinc alloys	0.05 to 0.08

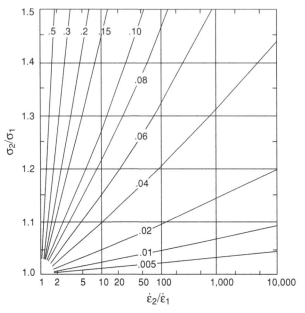

Figure 6.1. The dependence of flow stress on strain rate for several values of the strain-rate sensitivity, m, according to Equation 6.2. From W. F. Hosford and R. M. Caddell, *Metal Forming: Mechanics and Metallurgy*, 3rd Ed. Cambridge U. Press (2007).

Solution: Using Equation 6.2, $\dot{\varepsilon}_2/\dot{\varepsilon}_1 = (\sigma_2/\sigma_1)^{1/m} = (1.01)^{1/0.005} = 7.3$. If $m = 0.02$, $\dot{\varepsilon}_2/\dot{\varepsilon}_1 = 1.64$. The difference between the strain rates differently stressed locations decreases with increasing values of m.

Figure 6.2 illustrates two ways of determining the value of m. One method is to run two continuous tensile tests at different strain rates and compare the levels of stress at the some fixed strain. The other way is to change the strain rate suddenly during a test and compare the levels of stress immediately before and after the change. The latter method is easier and therefore more common. The two methods may give somewhat different values for m. In both cases,

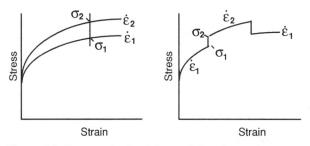

Figure 6.2. Two methods of determining the strain-rate sensitivity. Either continuous stress-strain curves at different strain rates can be compared at the same strain (left) or sudden changes of strain rate can be made and the stress levels just before and just after the change compared (right). In both cases, Equation 6.4 can be used to find m. In rate-change tests, $(\dot{\varepsilon}_2/\dot{\varepsilon}_1)$ is typically 10 or 100.

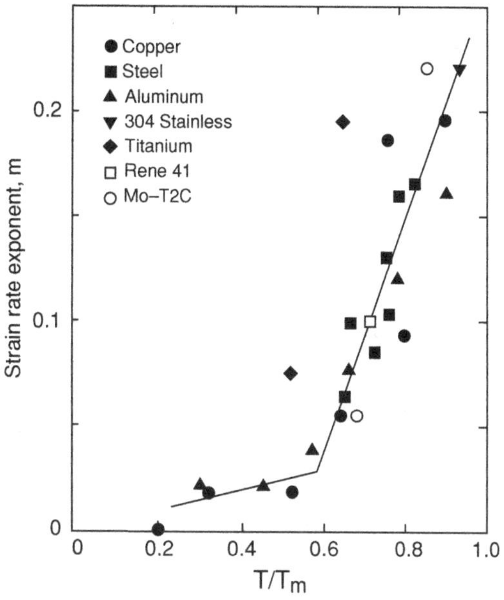

Figure 6.3. Variation of the strain-rate sensitivity, m, with temperature for several metals. Above about half of the melting point, m rises rapidly with temperature. From Hosford and Caddell, *ibid*.

Equation 6.4 can be used to find m. In rate-change tests, $(\dot{\varepsilon}_2/\dot{\varepsilon}_1)$ is typically 10 or 100.

For most metals, the value of the rate sensitivity, m, is low near room temperature but increases with increasing temperature. The increase of m with temperature is quite rapid above half the melting point $(T > T_m/2)$ on an absolute temperature scale. In some cases, m may be 0.5 or greater. Figure 6.3 shows the temperature dependence of m for several metals. For some alloys, there is a minimum between $0.2T_m$ and $0.3T_m$. For aluminum alloy 2024 (Figure 6.4), the rate sensitivity is slightly negative in this temperature range.

Superplasticity

If the rate sensitivity, m, of a material is 0.5 or greater, the material will behave *superplastically*, exhibiting very large tensile elongations. The large elongations occur because the necks are extremely gradual (like those that one observes when chewing gum is stretched). Superplasticity permits forming of parts that require very large strains. The conditions for superplasticity are:

1. Temperatures equal to or above half the absolute melting point $(T \geq 0.5T_m)$
2. Slow strain rates* (usually 10^{-3}/s or slower)
3. Very fine grain size (grain diameters of a few micrometers or less).

---

* Note that m is somewhat rate-dependent, so Equation 6.1 is not a complete description of the rate dependence. Nevertheless, m, defined by $m = \partial\ln\sigma/\partial\ln\dot{\varepsilon}$, is a useful index for describing the rate sensitivity of a material and analyzing rate effects.

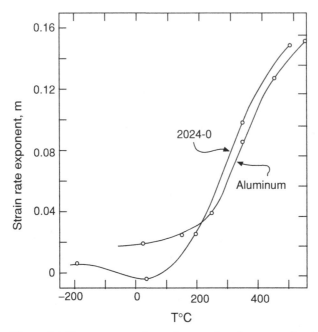

Figure 6.4. Temperature dependence of m for pure aluminum and aluminum alloy 2024. Note that at room temperature, m is negative for 2024. From Fields and Backofen, *Trans ASM*, v.51 (1959).

Long times are needed to form useful shapes at low strain rates. With long times and high temperatures grain growth may occur, negating an initially fine grain size. For this reason, most superplastic alloys either have two-phase structures or have a very fine dispersion of insoluble particles. Both minimize grain growth.

Under superplastic conditions, flow stresses are very low and extremely large elongations (1000% or more) are observed in tension tests. Both the low flow stress and large elongations can be useful in metal forming. The very low flow stresses permits slow forging of large, intricate parts with fine detail. The very large tensile elongation makes it possible to form very deep parts with simple tooling. Examples are shown in Figures 6.5 and 6.6.

The reason that very large tensile elongations are possible can be understood in terms of the behavior of a tensile bar having a region with a slightly reduced cross-sectional area (Figure 6.7). The following treatment is similar to the treatment in Chapter 4 of strain localization at defects. However, assume now that work-hardening can be neglected at high temperatures and that Equation 6.1 describes the strain-rate dependence. Let A_a and A_b be the cross-sectional areas of the thicker and thinner regions. The force carried by both sections is the same, so

$$F_b = \sigma_b A_b = F_a = \sigma_a A_a. \tag{6.6}$$

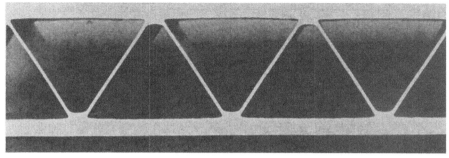

Figure 6.5. A titanium-alloy aircraft panel made by diffusion bonding and superplastic deformation. Three sheets were diffusion-bonded at a few locations and then internal pressurization of the unbonded channels caused the middle sheet to stretch. Superplastic behavior is required because of the very large uniform extension required in the middle sheet. From Hosford and Caddell, *ibid.*

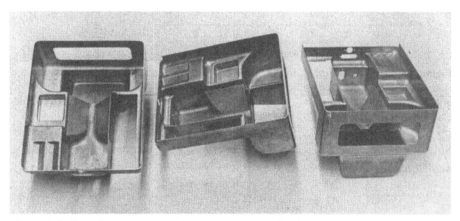

Figure 6.6. Complex part made from a sheet of a Zn-22%Al alloy. The very large strains in the walls require superplasticity. From Hosford and Caddell, *ibid.*

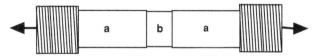

Figure 6.7. Schematic of a tensile bar with one region (b) slightly smaller in cross-section than the other (a). Both must carry the same tensile force.

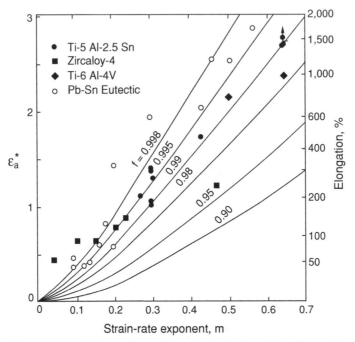

Figure 6.8. Limiting strain, ε_a^*, in the unnecked portion of a tensile specimen as a function of f and m. Reported values of total elongations are also shown as a function of m. These are plotted with the elongation converted to true strain. From Hosford and Caddell, *ibid.*

Substituting, $\sigma_a = C\dot{\varepsilon}_a^m$, $\sigma_b = C\dot{\varepsilon}_b^m$, $A_a = A_{ao}\exp(-\varepsilon_a)$, and $A_b = A_{bo}\exp(-\varepsilon_b)$, and Equation 6.6 becomes

$$A_{bo}\exp(-\varepsilon_b)C\dot{\varepsilon}_b^m = A_{ao}\exp(-\varepsilon_a)C\dot{\varepsilon}_a^m. \tag{6.7}$$

Designating the ratio of areas of the initial cross-sectional areas by $A_{bo}/A_{ao} = f$, $f \exp(-\varepsilon_b)\dot{\varepsilon}_b^m = \exp(-\varepsilon_a)\dot{\varepsilon}_a^m$. Now raising to the $(1/m)$th power and recognizing that $\dot{\varepsilon} = d\varepsilon/dt$, $f^{1/m}\exp(-\varepsilon_b/m)d\varepsilon_b = \exp(-\varepsilon_a/m)d\varepsilon_a$. Integration from zero to their current values, ε_b and e_a, results in

$$\exp(-\varepsilon_a/m) - 1 = f^{1/m}[\exp(-\varepsilon_b/m) - 1]. \tag{6.8}$$

Under superplastic conditions, the reduction in area is often quite large. The deformation in the unnecked region can be approximated by letting $\varepsilon_b \to \infty$ in Equation 6.8. In this case, the limiting strain in the unnecked region ε_a^* is

$$\varepsilon_a^* = -m\ln(1 - f^{1/m}). \tag{6.9}$$

Figure 6.8 shows how ε_a^* varies with m and f. Also shown are measured tensile elongations, plotted assuming that $\varepsilon_a^* = \ln(\%\text{El}/100 + 1)$. This plot suggests that values of f in the range from 0.99 to 0.995 are not unreasonable.

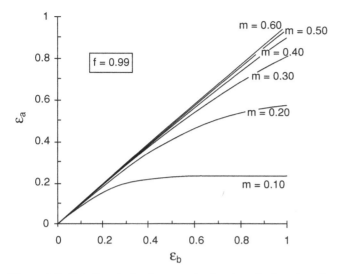

Figure 6.9. Relative strains in unnecked and necked regions for several values of rate sensitivity, m.

The large values of m are a result of the contribution of diffusional processes to the deformation. Two possible mechanisms are possible. One is a net diffusional flux under stress of atoms from grain boundaries parallel to the tensile axis to grain boundaries normal to the tensile axis, causing a tensile elongation. This amounts to a net diffusion of vacancies from grain boundaries normal to the tensile axis to boundaries parallel to it. The other possible mechanism is grain boundary sliding. With grain boundary sliding, compatibility where three grains meet must be accommodated by another mechanism. Such accommodation could occur either by slip or by diffusion, but in both cases the overall m would be below 1. Both diffusional creep and grain boundary sliding need a very fine grain size, high temperatures, and low strain rates. These mechanisms are discussed in more detail in Chapter 8.

EXAMPLE PROBLEM #6.3: A tensile bar is machined so that the diameter at one location is 1% smaller than the rest of the bar. The bar is tested at high temperature so strain hardening is negligible but the strain-rate exponent is 0.25. When the strain in the reduced region is 0.20, what is the strain in the larger region?

Solution: $f = (\pi D_1^2/4)/(\pi D_2^2/4) = (D_1/D_2)^2 = (0.99/1)^2 = 0.98$. Substituting $\varepsilon_b = 0.20$ and $f = 0.98$ into Equation 6.8, $\exp(-\varepsilon_a/m) - 1 = f^{1/m}[\exp(-\varepsilon_a/m) - 1]$, $\exp(-\varepsilon_a/0.25) = (0.98)^{1/0.25}[\exp(-0.2/0.25) - 1] + 1 = 00.492$. $\varepsilon_a = -0.25\ln(0.492) = 0.177$.

Figure 6.9 shows that for large values of m, the strain, ε_a, in the region outside of the neck continues to grow even when the neck strain, ε_b, is large. With low values of m, deformation outside of the neck ceases early.

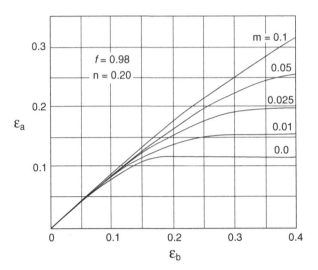

Figure 6.10. Comparison of the strains in the reduced and unreduced sections of a tensile bar for a material that strain hardens and is strain-rate sensitive. From Hosford and Caddell, *ibid*.

Combined Strain and Strain-Rate Effects

Rate sensitivity may have appreciable effects at low temperatures, even when m is relatively small. If both strain and strain-rate hardening are considered, the true stress may be approximated by

$$\sigma = C'\varepsilon^n\dot{\varepsilon}^m. \tag{6.10}$$

Reconsider the tension test on an inhomogeneous tensile bar, with initial cross-sections of A_{ao} and $A_{bo} = fA_{oa}$. Substituting $A_a = A_{ao}\exp(-\varepsilon_a)$, $A_b = A_{bo}\exp(-\varepsilon_b)$, $\sigma_a = C\varepsilon_a^n e_a \dot{\varepsilon}^m$ and $\sigma_b = C\varepsilon_b^n \dot{\varepsilon}_b e^m$ into a force balance (Equation 6.6) $F_a = A_a\sigma_a = F_b = A_b\sigma_b$,

$$A_{ao}\exp(-\varepsilon_a)C'\varepsilon_a^n\dot{\varepsilon}_a^m = A_{bo}\exp(-\varepsilon_b)C'\varepsilon_b^n\dot{\varepsilon}_b^m. \tag{6.11}$$

Following the procedure leading to Equation 6.8,

$$\int_o^{\varepsilon a} \exp(-\varepsilon_a/m)\varepsilon_a^{n/m}\mathrm{d}\varepsilon_a = f^{1/m}\int_o^{\varepsilon b} \exp(-\varepsilon_b/m)\varepsilon_b^{n/m}\mathrm{d}\varepsilon_b. \tag{6.12}$$

This equation must be integrated numerically. The results are shown in Figure 6.10 for $n = 0.2$, $f = 0.98$, and several levels of m.

EXAMPLE PROBLEM #6.4: Reconsider the tension test in example Problem #5.3. Let $n = 0.226$ and $m = 0.012$ (typical values for low-carbon steel) and let $f = 0.98$ as in the previous problem. Calculate the strain, ε_a, in the region with the larger diameter if the bar necks to a 25% reduction of area, $\varepsilon_b = \ln(4/3) = 0.288$.

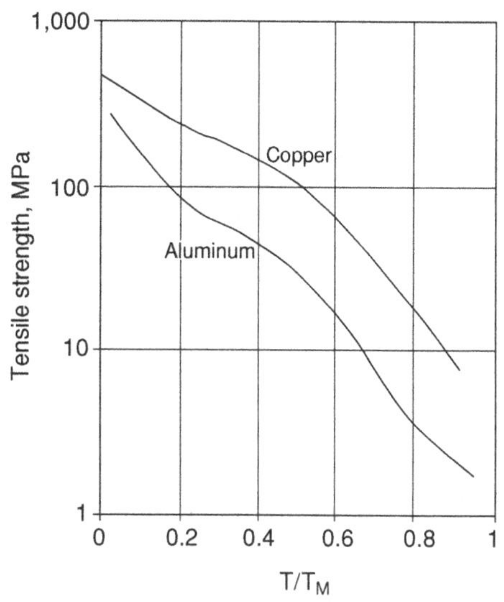

Figure 6.11. Tensile strengths decrease with increasing temperatures. Data from R. P. Carrecker and W. R. Hibbard Jr., *Trans. TMS-AIME*, v. 209 (1957).

Solution: Substituting into Equation 6.12,

$$\int_{o}^{\varepsilon_a} \exp(-\varepsilon_a/m)\varepsilon_a^{n/m}\mathrm{d}\varepsilon_a = f^{1/m}\int_{o}^{\varepsilon_b} \exp(-\varepsilon_b/m)\varepsilon_b^{n/m}\mathrm{d}\varepsilon_b$$

or

$$\int_{o}^{\varepsilon_a} \exp(-83.33\varepsilon_a)\varepsilon_a^{18.83}\mathrm{d}\varepsilon_a = 0.1587\int_{o}^{\varepsilon_b} \exp(-83.33\varepsilon_b)\varepsilon_b^{18.83}\mathrm{d}\varepsilon_b$$

Numerical integration of the right-hand side gives 0.02185. The left-hand side has the same value when $\varepsilon_a = 0.2625$. The loss of elongation in the thicker section is much less than in example Problem 6.3 where it was found that $\varepsilon_a = 0.195$ with $n = 0.226$ and $m = 0$.

Temperature Dependence

As temperature increases, the whole level of the stress-strain curve generally drops. Figure 6.11 shows the decrease of tensile strengths of copper and aluminum. Usually, the rate of work-hardening also decreases at high temperatures.

Combined Temperature and Strain-Rate Effects

The effects of temperature and strain rate are interrelated. Decreasing temperature has the same effect as an increasing strain rate, as shown schematically in Figure 6.12. This effect occurs even in temperature regimes where the rate sensitivity is negative.

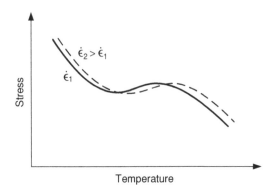

Figure 6.12. Schematic plot showing the temperature dependence of flow stress at two different strain rates. Note that an increased strain rate has the same effects as a decreased temperature. From Hosford and Caddell, *ibid.*

The simplest treatment of temperature dependence of strain rate is that of Zener and Hollomon,* who treated plastic straining as a thermally activated rate process. They assumed that strain rate follows an Arrhenius rate law, rate $\alpha \exp(-Q/RT)$, that is widely used in analyzing many temperature-dependent rate processes. They proposed

$$\dot{\varepsilon} = A\exp(-Q/RT), \tag{6.13}$$

where Q is the activation energy, T is absolute temperature, and R is the gas constant. At a fixed strain, A is a function only of stress, $A = A(\sigma)$, so Equation 6.13 can be written as

$$A(\sigma) = \dot{\varepsilon}\exp(+Q/RT) \tag{6.14}$$

or

$$A(\sigma) = Z, \tag{6.15}$$

where $Z = \dot{\varepsilon}\exp(+Q/RT)$ is called the Zener-Hollomon parameter.

Equation 6.20 predicts that a plot of strain at constant stress on a logarithmic scale vs. $1/T$ should be linear. Figure 6.13 shows that such a plot for aluminum alloy 2024 is linear for a wide range of strain rates although the relation fails at large strain rates.

EXAMPLE PROBLEM #6.6: For an aluminum alloy under constant load, the creep rate increased by a factor of two when the temperature was increased from 71°C to 77°C. Find the activation energy.

Solution: Because $\dot{\varepsilon} = A\exp(-Q/RT)$, the ratio of the creep rates at two temperatures is $\dot{\varepsilon}_2/\dot{\varepsilon}_1 = \exp[(-Q/R)(1/T_2 - 1/T_1)]$ so $Q = R\ln(\dot{\varepsilon}_2/\dot{\varepsilon}_1)/(1/T_1 - 1/T_2)$. Substituting $T_1 = 273 + 71 = 344K$, $T_2 = 273 + 77 = 350K$, $R = 8.314$ J/mole/K and $\dot{\varepsilon}_2/\dot{\varepsilon}_1 = 2$ and solving, $Q = 116$ kJ/mol. Compare this with Figure 6.14.

* C. Zener and H. Hollomon, *J. Appl. Phys.* 15 (1944).

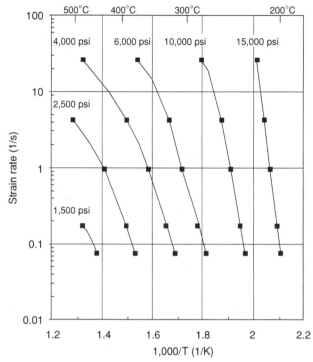

Figure 6.13. Combinations of strain rate and temperature for several stresses in aluminum alloy 2024. The data are taken at an effective strain of 1.0. Data from Fields and Backofen, *ibid*.

The Zener-Hollomon development is useful if the temperature and strain-rate ranges are not too large. Dorn and coworkers measured the activation energy, Q, for pure aluminum over a very large temperature range. They did this by observing the change of creep rate under fixed load when the temperature was suddenly changed. Because the stress is constant,

$$\dot{\varepsilon}_2/\dot{\varepsilon}_1 = \exp[(-Q/R)(1/T_2 - 1/T_1)] \qquad (6.16)$$

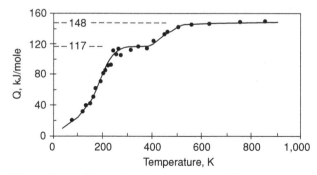

Figure 6.14. Measured activation energies for creep of aluminum as a function of temperature. The change of Q with temperature indicates that Equation 6.20 will lead to errors if it is applied over a temperature range in which Q changes. Data from O. D. Sherby, J.L. Lytton and J.E. Dorn, *AIME Trans.* v. 212 (1958).

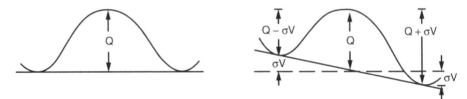

Figure 6.15. Schematic illustration of the skewing of an activation barrier by an applied stress. From Hosford and Caddell, *ibid.*

or

$$Q = \mathrm{R}\ln(\dot{\varepsilon}_2/\dot{\varepsilon}_1)/(1/T_1 - 1/T_2). \tag{6.17}$$

Their results (Figure 6.14) indicated that Q was independent of temperature above 500K but very temperature-dependent at lower temperatures. This observation was later explained by Z. S. Basinski and others. The basic argument is that the stress helps thermal fluctuations overcome activation barriers. This is illustrated in Figure 6.15. If there is no stress, the activation barrier has a height of Q with random fluctuations. The rate of overcoming the barrier is proportional to $\exp(-Q/RT)$. An applied stress skews the barrier so that the effective height of the barrier is reduced to $Q - \sigma v$, where v is a parameter with the units of volume. Now the rate of overcoming the barrier is proportional to $\exp[-(Q - \sigma v)/RT]$. The finding of lower activation energies at lower temperatures was explained by the fact that in the experiments, greater stresses were applied at lower temperatures to achieve measurable creep rates.

In Figure 6.16, the rate of the overcoming the barrier from left to right is proportional to $\exp[-(Q - \sigma v)/RT]$, whereas the rate from right to left is proportional to $\exp[-(Q + \sigma v)/RT]$. Thus, the net reaction rate is $C\{\exp[-(Q - \sigma v)/RT] - \exp[-(Q + \sigma v)/RT]\} = C\exp(-Q/RT)\{\exp[(\sigma v)/RT] - \exp[-(\sigma v)/RT]\}$. This simplifies to

$$\dot{\varepsilon} = 2C\exp(-Q/RT)\sinh[(\sigma v)/RT]. \tag{6.18}$$

This expression has been modified based on some theoretical arguments to provide a better fit with experimental data,

$$\dot{\varepsilon} = A\exp(-Q/RT)[\sinh(\alpha\sigma)]^{1/m}, \tag{6.19}$$

where a is an empirical constant and the exponent $1/m$ is consistent with Equation 6.1. Figure 6.16 shows that Equation 6.19 can be used to correlate the combined effects of temperature, stress, and strain rate over an extremely large range of strain rates.

If $\alpha\sigma \ll 1$, $\sinh(\alpha\sigma) \approx \alpha\sigma$, Equation 6.19 simplifies to

$$\dot{\varepsilon} = A\exp(+Q/RT)(\alpha\sigma)^{1/m} \tag{6.20}$$

$$\sigma = \dot{\varepsilon}^m A'\exp(-mQ/RT) \tag{6.21}$$

$$\sigma = A'Z^m \tag{6.22}$$

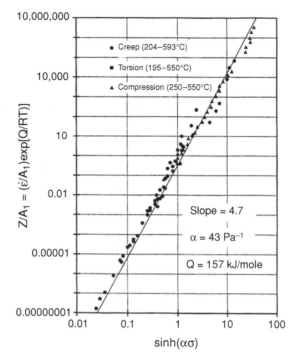

Figure 6.16. Plot of the Zener-Hollomon parameter vs. flow-stress data for aluminum over a very wide range of strain rates. Data are from J. J. Jonas, *Trans. Q. ASM*, v. 62 (1969).

where $A' = (\alpha A^m)^{•1}$. Equation 6.22 is consistent with the Zener-Hollomon development. At low temperatures and large stresses, $(\alpha\sigma) \gg 1$, so $\sinh(\alpha\sigma)$ $\rightarrow \exp(\alpha\sigma)/2$ and Equation 6.20 reduces to

$$\dot\varepsilon = C\exp(\alpha'\sigma - Q/RT). \tag{6.23}$$

Under these conditions, both C and α' depend on strain and temperature. For a constant temperature and strain,

$$\sigma = C + m'\ln\dot\varepsilon \tag{6.24}$$

Note that this equation differs from Equation 6.1.

Hot Working

Metallurgists make a distinction between *cold working* and *hot working*. It is often said that hot working is done above the recrystallization temperature, and the work material recrystallizes as it is deformed. This is an oversimplification. The strain rates during hot working are often so large that there is not enough time for recrystallization during the deformation. It is more meaningful to think of hot working as a process in which recrystallization occurs in the period between repeated operations as in forging, rolling, and so on, or at least before the work material cools from the working temperature to room temperature.

Tool forces during hot working are lowered not only by the recrystallization itself but also because of the inherently lower flow stresses at high temperatures. A second advantage of hot working is that the resulting product is in an annealed state. However, there are disadvantages of hot working. Among these are:

1. The tendency of the work metal to oxidize. In the case of steel and copper-base alloys, this results in scale formation and a consequent loss of metal. Titanium alloys must be hot worked under an inert atmosphere to avoid dissolving oxygen.
2. Lubrication is much more difficult, and consequently friction much higher. This effect somewhat negates the effect of lower flow stress.
3. The scaling and high friction result in tool wear and shortened tool life.
4. The resulting product surface is rough because of the scale, tool wear, and poor lubrication. There is also a loss of precise gauge control.
5. In some cases, the lack of strain hardening in hot working is undesirable.

Usually, ingot breakdown and subsequent working are done hot until the section size becomes small enough that high friction and the loss of material by oxidation become important. Hot rolling of steel sheet is continued to a thickness of 0.050 in. to 0.200 in. Then the product is pickled in acid to remove the scale and finished by cold working to obtain a good surface.

Most cold-rolled sheet is sold in an annealed condition. With steels, the term *cold rolled* refers to the surface quality rather than the state of work-hardening. The term *warm working* is used to refer to working of metals above room temperature but at a low enough temperature that recrystallization does not occur.

REFERENCES

Metal Forming: Mechanics and Metallurgy: 2nd ed., W. F. Hosford and R. M. Caddell, Cambridge U. Press (2007).
Ferrous Metallurgical Design, John H. Hollomon and Leonard Jaffe, John Wiley & Sons (1947).
W. F. Hosford, *Mechanical Behavior of Materials*, Cambridge U. Press, (2005).

Notes

Svante August Arrhenius (1859–1927) was a Swedish physical chemist. He studied at Uppsala, where he obtained his doctorate in 1884. It is noteworthy that his thesis on electrolytes was given a fourth (lowest) level pass because his committee was skeptical of its validity. From 1886 to 1890, he worked with several noted scientists in Germany who did appreciate his work. In 1887, he suggested that a very wide range of rate processes follow what is now known as the Arrhenius equation. For years, his work was recognized throughout the world except in his native Sweden.

Count Rumford (Benjamin Thompson) was the first person to measure the mechanical equivalent of heat. He was born in Woburn, Massachusetts, in 1753, studied at Harvard and taught in Concord, New Hampshire, where he married a wealthy woman. At the outbreak of the American Revolution, having been denied a commission by Washington, he approached the British who did commission him. When the British evacuated Boston in 1776 he left for England, where he made a number of experiments on heat. However, he was forced to depart quickly for Bavaria after being suspected of selling British naval secrets to the French. In the Bavarian army, he rose rapidly to become Minister of War and eventually Prime Minister. While inspecting a canon factory, he observed a large increase in temperature during the machining of bronze canons. He measured the temperature rise and with the known heat capacity of the bronze, he calculated the heat generated by machining. By equating this to the mechanical work done in machining, he was able to deduce the mechanical equivalent of heat. His value was a little too low, in part due to the fact some of the plastic work done in machining is stored as dislocations in the chips.

Thompson was knighted in 1791, choosing the title Count Rumford after the original name of Concord. During his rapid rise to power in Bavaria he made many enemies, so that when the winds of politics changed, he had to leave. He returned to Britain as an ambassador from Bavaria, but the British refused to recognize him. During this period, he established the Royal Institute. At this time, the United States was establishing the U.S. Military Academy at West Point, and Rumford applied for and was selected to be its first commandant. However, during the negotiations about his title, it was realized that he had deserted the American cause during the Revolution and the offer was withdrawn. He finally moved to Paris where he married the widow of the famous French chemist, Lavousier, who had been beheaded during the French revolution.

Problems

1. During a tension test, the rate of straining was suddenly doubled. This caused the load (force) to rise by 1.2%. Assuming that the strain-rate dependence can be described by $\sigma = C\dot{\varepsilon}^m$, what is the value of m?

2. Two tension tests were made on the same alloy but at different strain rates. Both curves were fitted to a power-law strain hardening expression of the form, $\sigma = K\varepsilon^n$. The results are summarized in Table 6.2. Assuming that the flow stress at constant strain can be approximated by $\sigma = C\dot{\varepsilon}^m$, determine the value of m.

Table 6.2. *Results of two tension tests*

	Test A	Test B
strain rate (s^{-1})	2×10^{-3}	10^{-1}
strain-hardening exponent, n	0.22	0.22
constant, K (MPa)	402	412

3. To achieve a weight-saving in an automobile, replacement of a low-carbon steel with an HSLA steel is considered. In laboratory tension tests at a strain rate of 10^{-3}/s, the yield strengths of the HSLA steel and the low-carbon steel were measured to be 400 MPa and 220 MPa, respectively. The strain-rate exponents are $m = 0.005$ for the HSLA steel and $m = 0.015$ for the low-carbon steel. What percent of weight-saving could be achieved if the substitution was made so that the forces were the same at the strain rates of 10^{+3}, typical of crash conditions?

4. The thickness of a cold-rolled sheet varies from 0.0322 in. to 0.0318 in., depending on where the measurement is made, so strip tensile specimens cut from the sheet show similar variation in cross-section.
   A. For a material with $n = 0.20$ and $m = 0$, what will be the thickness of the thickest regions when the thinnest region necks?
   B. Find the strains in the thicker region if $m = 0.50$ and $n = 0$ when the strain in the thinnest region reaches
      i. 0.5,   ii. ∞

5. Estimate the total elongation of a superplastic material if
   A. $n = 0, m = 0.5$, and $f = 0.98$,   B. $n = 0, f = 0.75$, and $m = 0.8$.

6. In superplastic forming, it is often necessary to control the strain rate. Consider the forming of a hemispherical dome by clamping a sheet over a circular hole and bulging it with gas pressure.
   A. Compare gas pressure needed to form a 2.0-in. dome with that needed to form a 20-in. dome if both are formed from sheets of the same thickness and at the same strain rate.
   B. Describe (qualitatively) how the gas pressure should be varied during the forming to maintain a constant strain rate.

7. During a constant-load creep experiment on a polymer, the temperature was suddenly increased from 100°C to 105°C. It was found that this increase of temperature caused the strain rate to increase by a factor of 1.8. What is the apparent activation energy for creep of the plastic?

8. Figure 6.17 gives some data for the effect of stress and temperature on the strain rate of a nickel-based super-alloy single crystal. The strain rate is independent of strain in the region of this data. Determine as accurately as possible:
   A. The activation energy, Q, in the temperature range 700°C to 810°C.
   B. The exponent, m, for 780°C.

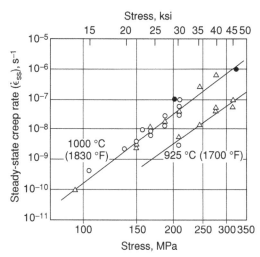

Figure 6.17. The effects of stress and temperature on the strain rate of nickel–based super-alloy single crystals under stress. From *Metals Hanbdbook*, 9th ed., v.8, ASM (1985).

9. The stress-strain curve of a steel is represented by $\sigma = (1600\,\text{MPa})\varepsilon^{0.1}$. The steel is deformed to a strain of $\varepsilon = 0.1$ under adiabatic conditions. Estimate the temperature rise in the sample. For iron, $\rho = 7.87\,\text{kg/m}^2$, $C = 0.46\text{J/g}°\text{C}$, and $E = 205\,\text{GPa}$.

10. Evaluate m for copper at room temperature from Figure 6.13.

11. The stress-strain curves for silver at several temperatures and strain rates are shown in Figure 6.18. Determine the strain-rate exponent, m, for silver at 25°C

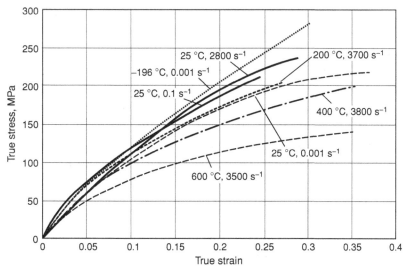

Figure 6.18. Stress-strain curves for silver (fcc). Note that at 25°C the difference, $\Delta\sigma$, between the stress-strain curves at different rates is proportional to the stress level. From G. T. Gray in *ASM Metals Handbook*, v. 8 (2000).

7 Viscoelasticity

Introduction

In classic elasticity, there is no time delay between application of a force and the deformation that it causes. However, for many materials there is additional time-dependent deformation that is recoverable. This is called *viscoelastic* or *anelastic* deformation. When a load is applied to a material, there is an instantaneous elastic response but the deformation also increases with time. This viscoelasticity should not be confused with *creep* (Chapter 8), which is time-dependent plastic deformation. Anelastic strains in metals and ceramics are usually so small that they are ignored. However, in many polymers viscoelastic strains can be very significant.

Anelasticity is responsible for the damping of vibrations. A large damping capacity is desirable where vibrations might interfere with the precision of instruments or machinery and for controlling unwanted noise. A low damping capacity is desirable in materials used for frequency standards, such as in bells and in many musical instruments. Viscoelastic strains are often undesirable. They cause the sagging of wooden beams, denting of vinyl flooring by heavy furniture, and loss of dimensional stability in gauging equipment. The energy associated with damping is released as heat, which often causes an unwanted temperature increase. The study of damping peaks and how they are affected by processing has been useful in identifying mechanisms. The mathematical descriptions of viscoelasticity and damping will be developed in the first part of this chapter, then several damping mechanisms will be described.

Rheological Models

Anelastic behavior can be modeled mathematically with structures constructed from idealized elements representing elastic and viscous behavior, as shown in Figure 7.1.

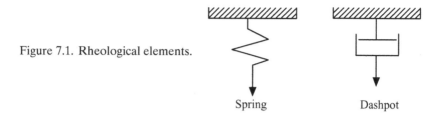

Figure 7.1. Rheological elements.

Spring   Dashpot

A spring models a perfectly elastic solid. The behavior is described by

$$e_e = F_e/K_e, \tag{7.1}$$

where e_e is the change of length of the spring, F_e is the force on the spring, and K_e is the spring constant. A dashpot models a perfectly viscous material. Its behavior is described by

$$\dot{e}_v = de_v/dt = F_v/K_v, \tag{7.2}$$

where e_v is the change in length of the dashpot, F_v is the force on it, and K_v is the dashpot constant.

Series Combination of a Spring and Dashpot

The *Maxwell model* consists of a spring and dashpot in series as shown in Figure 7.2. Here and in the following, e and F, without subscripts, will refer to the overall elongation and the external force. Consider how this model behaves in two simple experiments. First, let there be a sudden application of a force, F, at time, $t = 0$, with the force being maintained constant (Figure 7.3). The immediate response from the spring is $e_e = F/K_e$. This is followed by a time-dependent response from the dashpot, $e_v = Ft/K_v$. The overall response will be

$$E = e_e + e_v = F/K_e + Ft/K_v, \tag{7.3}$$

so the strain rate would be constant. The viscous strain would not be recovered on unloading.

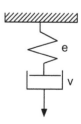

Figure 7.2. Spring and dashpot in series (Maxwell model).

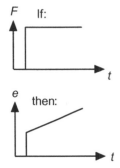

Figure 7.3. Strain relaxation predicted by the series model. The strain increases linearly with time.

Now consider a second experiment. Assume that the material is forced to undergo a sudden elongation, e, at time $t = 0$, and that this elongation is maintained for period of time as sketched in Figure 7.4. Initially, the elongation must be accommodated entirely by the spring ($e = e_e$), so the force initially jumps to a level, $F_o = K_e e$. This force causes the dashpot to operate, gradually increasing the strain e_v. The force in the spring, $F = K_e e_e = (e - e_v)K_e$, equals the force in the dashpot, $F = K_v de_v/dt$, $(e - e_v)^{-1}de_v = -(K_e/K_v)dt$. Integrating, $\ln[(e-e_v)/e] = -(K_e/K_v)t$. Substituting, $(e-e_v) = F/K_e$ and $K_e e = F_o$, $\ln(F/F_o) = t/(K_v/K_e)$. Now defining a relaxation time, $\tau = K_v/K_e$,

$$F = F_o \exp(-t/\tau). \tag{7.4}$$

Parallel Combination of Spring and Dashpot

The *Voight model* consists of a spring and dashpot in parallel, as sketched in Figure 7.5. For this model, $F = F_e + F_v$ and $e = e_e = e_v$.

Now consider the behavior of the Voight model in the same two experiments. In the first, there is sudden application at time, $t = 0$, of a force, F, which is then maintained at that level (Figure 7.6). Initially, the dashpot must carry the entire force because the spring can carry a force only when it is extended. At an infinite time, the spring carries all the force, so $e_\infty = F/K_e$. Substituting $de_v = de$ and $F_v = F - F_e$ into $de_v/dt = F_v/K$, $de/dt = (F - F_e)/K_v$. Now substituting $F_e = e_e K_e = e K_e$, $de/dt = (F - e K_e)/K_v = (K_e/K_v)(F/K_e - e)$. Denoting F/K_e by e_∞, defining the relaxation time as

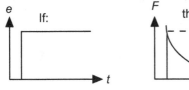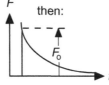

Figure 7.4. Stress relaxation predicted by the series model. The stress decays to zero.

Figure 7.5. Spring and dashpot in parallel (Voight model).

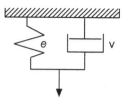

$\tau = K_v/K_e$ and rearranging, $\int(e_\infty - e)^{-1}de = \int dt/t$. Integrating, $\ln[(e - e_\infty)/-e_\infty] = t/\tau$. Rearranging, $(e_\infty - e)/e_\infty = \exp(-t/t)$ or

$$e = e_\infty[1 - \exp(-t/\tau)]. \tag{7.5}$$

Note that $e = 0$ at $t = 0$, and $e \to e_\infty$ as $t \to \infty$.

The experiment in which an extension is suddenly applied to the system (Figure 7.7) is impossible because the dashpot cannot undergo an sudden extension without an infinite force.

Combined Parallel-Series Model

Neither the series nor the parallel model adequately describes both stress and strain relaxation. A combined series-parallel or *Voight-Maxwell model* is much better. In Figure 7.8, spring #2 is in parallel with a dashpot and spring #1 is in series with the combination. The basic equations of this model are $F = F_1 = F_2 + F_v$, $e_v = e_2$, and $e = e_1 = e_2$.

Consider first the sudden application of a force, F, at time $t=0$ (Figure 7.9). One can write $de_v/dt = F_v/K_v = (F - F_2)/K_v = (F - K_2e_2)/K_v = (F/K_2 - e_v)(K_2/K_v)$. Rearranging, $\int(F/K_2 - e_v)^{-1}de_v = (1/\tau_e)\int dt$, where $\tau_e = K_v/K_2$ is the relaxation time for strain relaxation. Integration gives $\ln[(F/K_2 - e_v)/F/K_2)] = -t/\tau_e$ or $e_v = (1/K_2)F[1 - \exp(-t/\tau_e)]$. Substituting $F/K_2 = e_\infty - e_o$ where e_∞ and e_o are the relaxed and initial (unrelaxed) elongations, $e_v = (e_\infty - e_o)(1 - \exp(-t/\tau_e)$. The total strain is $e = e_v + e_1$, so

$$e = e_\infty - (e_\infty - e_o)\exp(-t/\tau_e). \tag{7.6}$$

Note that $e = e_o$ at $t = 0$ and $e \to e_\infty$ as $t \to \infty$.

Now consider the experiment in which an elongation, e, is suddenly imposed on the material. Immediately after stretching, all of the strain occurs in spring #1, so $e = F/K_1$, so the initial force, $F_o = eK_1$ (Figure 7.10). After an infinite time, the dashpot carries no load, so $e = (1/K_1 + 1/K_2)F$, or

$$F_\infty = [K_1K_2/(K_1+K_2)]e. \tag{7.7}$$

Figure 7.6. Strain relaxation predicted by the parallel model. The strain saturates at e_∞.

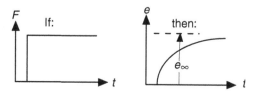

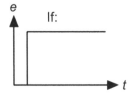

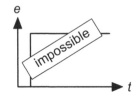

Figure 7.7. An instantaneous strain cannot be imposed on the parallel model.

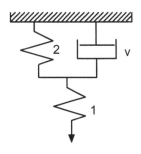

Figure 7.8. Combined series-parallel model (Voight-Maxwell model).

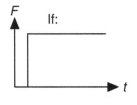

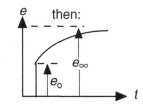

Figure 7.9. Strain relaxation predicted by the series-parallel model. The strain saturates at e_∞.

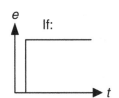

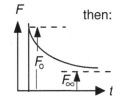

Figure 7.10. Stress relaxation with the series-parallel model. The stress decays to F_∞.

The force decaying from F_o to F_∞ is given by

$$F = F_o - (F_o - F_\infty)\exp(-t/\tau_\sigma), \tag{7.8}$$

where the relaxation time, τ_σ, is given by

$$\tau_\sigma = K_v/(K_1 + K_2). \tag{7.9}$$

Note that the relaxation time for stress relaxation is shorter than the relaxation time for strain relaxation,

$$\tau_e = K_v/K_2. \tag{7.10}$$

EXAMPLE PROBLEM #7.1: For a given material, $F_o/F_\infty = 1.02$. Find the ratio of τ_e/τ_σ for this material.

Solution: Combining Equations 15.9 and 15.10, $\tau_e/\tau_\sigma = (K_1 + K_2)/K_2$, For strain relaxation, $e = F_o/K_1 = F_\infty(1/K_1 + 1/K_2)$, so $F_o/F_\infty = K_1(1/K_1 + 1/K_2) = (K_1 + K_2)/K_2$. Therefore $\tau_e/\tau_\sigma = F_o/F_\infty = 1.02$.

More Complex Models

More complicated models may be constructed using more spring and dashpot elements or elements with nonlinear behavior. Nonlinear elasticity can be modeled by a nonlinear spring for which $F = K_e f(e)$. Non-Newtonian viscous behavior can be modeled by a nonlinear dashpot for which $F = K_v f(\dot{e})$.

Damping

Viscoelastic straining causes damping. Consider the cyclic loading of a viscoelastic material,

$$\sigma = \sigma_o \sin(wt). \tag{7.11}$$

The strain,

$$e = e_o \sin(wt - \delta), \tag{7.12}$$

lags the stress by δ, as shown in Figure 7.11. A plot of stress vs. strain is an ellipse. The rate of energy loss per volume is given by $dU/dt = \sigma \, de/dt$, so the energy loss per cycle per volume, ΔU, is

$$\Delta U = \oint \sigma \, d\varepsilon, \tag{7.13}$$

or $\Delta U = \int \sigma[d\varepsilon/d(\omega t)]d(\omega t)$, where the integration limits are 0 and 2π. Substituting $\sigma = \sigma_o \sin(\omega t)$ and $d\varepsilon/d(\omega t) = \varepsilon_o \cos(\omega t - \delta)$, $\varepsilon = \varepsilon_o \sin(\omega t - \delta)$ $\Delta U = \sigma_o \varepsilon_o \int \sin(\omega t)\cos(\omega t - \delta)d(\omega t)$. But $\cos(\omega t - \delta) = \cos(\omega t)\cos\delta + \sin(\omega t)\sin\delta$, so

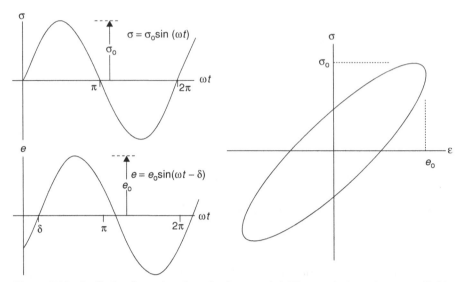

Figure 7.11. Cyclic loading of a viscoelastic material. The strain lags the stress (left). The plot of stress vs. strain is an ellipse (right).

$$\Delta U/(\sigma_o \varepsilon_o) = \int \cos\delta \cos(\omega t)\sin(wt)\mathrm{d}(\omega t) + \int \sin\delta \sin^2(\omega t)\mathrm{d}(\omega t) = \pi \sin\delta, \text{ or}$$

$$\Delta U = \pi(\sigma_o \varepsilon_o)\sin\delta. \qquad (7.14)$$

Because the elastic energy per volume required to load the material to σ_o and ε_o is $u = (1/2)\sigma_o \varepsilon_o$, this can be expressed as

$$\Delta U/U = 2\pi \sin\delta. \qquad (7.15)$$

If δ is small,

$$\Delta U/U = 2\pi \delta. \qquad (7.16)$$

Natural Decay

During free oscillation, the amplitude will gradually decrease, as shown in Figure 7.12. It is usually assumed that the decrease between two successive cycles is proportional to the amplitude, e. A commonly used measure of damping is the logarithmic decrement, Λ, defined as

$$\Lambda = \ln(e_n/e_{n}+_1). \qquad (7.17)$$

Λ can be related to δ by recalling that $U_n = \sigma_n e_n/2 = E e_n^2/2$, so $e_n = [2u_n/E]^{1/2}$ and $e_{n+1} = [2U_{n+1}/E]^{1/2}$. Substituting, $\Lambda = \ln\{[2U_{n+1}/E]/[2U_n/E]\}^{1/2} = \ln(U_{n+1}/U_n)^{1/2} = \ln(1+\Delta U/U)^{1/2}$, or

$$\Lambda = (1/2)\ln(1+\Delta U/U). \qquad (7.18)$$

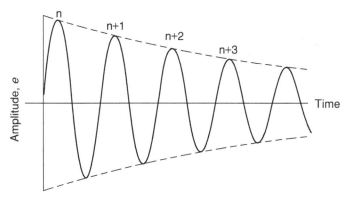

Figure 7.12. Decay of a natural vibration.

Using the series expansion, $\Lambda = (1/2)[\Delta U/U - (\Delta U/U)^2/2 + (\Delta U/U)^3/3 - \cdots$

For small values of $\Delta U/U$, $\delta \approx (\Delta U/U)/2$, or

$$\Lambda \approx \pi\delta. \tag{7.19}$$

Sometimes the extent of damping is denoted by Q^{-1}, where $Q^{-1} = \tan\Lambda \approx \pi\delta$.

EXAMPLE PROBLEM #7.2: The amplitude of a vibrating member decreases so that the amplitude on the 100th cycle is 13% of the amplitude on the first cycle. Determine Λ.

Solution: $\Lambda = \ln(e_n/e_{n+1})$. Because $(e_m/e_{m+1}) = (e_n/e_{n+1}) = (e_{n+1}/e_{n+2})$ and so on and $(e_n/e_{n+m}) = (e_n/e_{n+1})(e_{n+1}/e_{n+2})(e_{n+2}/e_{n+3})\ldots(e_{n+m-1}/e_{n+m}) = \mathrm{m}(e_n/e_{n+m})$.

Therefore, $(e_n/e_{n+1}) = (e_n/e_{n+m})^{1/m}$. $\Lambda = \ln(e_n/e_{n+1}) = \ln(e_n/e_{n+m})^{1/m} = (1/m)\ln(e_n/e_{n+m})$. Substituting $m = 100$ and $(e_n/e_{n+m}) = 1/0.13$, $\Lambda = 0.0204$.

Elastic Modulus – Relaxed vs. Unrelaxed

If the frequency is large enough, there is no relaxation and therefore no damping and the modulus, $E = \sigma/e$, is high. At low frequencies, there is complete relaxation with the result that there is no damping but the modulus will be lower. At intermediate frequencies, there is partial relaxation with high damping and a frequency-dependent modulus. The frequency dependence of the modulus is given by

$$E/E_r = \omega^2(\tau_e - \tau_\sigma)/(1 + \omega^2\tau_e^2). \tag{7.20}$$

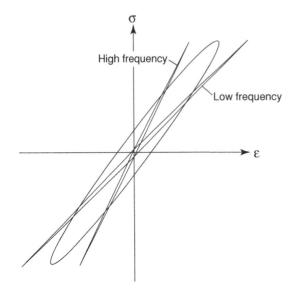

Figure 7.13. Hysteresis curves at a high, low, and intermediate frequency. The damping is low at both high and low frequencies.

The hysteresis curves for high, low, and intermediate frequencies are illustrated in Figure 7.13 and Figure 7.14 illustrates the corresponding interdependence of damping and modulus.

Thermoelastic Effect

When a material is elastically deformed rapidly (adiabatically), it undergoes a temperature change,

$$\Delta T = -\sigma \alpha T / C_v, \tag{7.21}$$

where T is the absolute temperature, α is the linear coefficient of thermal expansion, and C_v is the volume heat capacity. For most materials, elastic

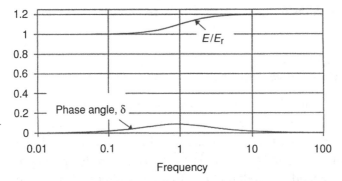

Figure 7.14. Dependence of damping and elastic modulus on frequency. The peak damping occurs at the frequency that the modulus is rapidly changing. E/E_r was calculated from Equation 7.20 with $\tau_e = 1.2\tau_\sigma$ and δ was calculated from $\tan\delta = \omega(\tau_e - \tau_\sigma)/(1 + \omega^2 \tau_e \tau_\sigma)$.

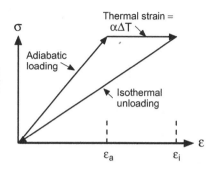

Figure 7.15. Schematic of loading adiabatically, holding until thermal equilibrium is reached, and then unloading isothermally.

stretching leads to a cooling because all the terms are positive. Rubber is an exception because α is negative when it is under tension.

Consider a simple experiment illustrated by Figure 7.15. Let the material be elastically loaded in tension adiabatically. It undergoes an elastic strain, $e_a = \sigma/E_a$, where E_a is the Young's modulus under adiabatic conditions. At the same time, it will experience a cooling, $\Delta T = -\sigma\alpha T/C_v$. Now let it warm back up to ambient temperature while still under the stress, σ, changing its temperature by $\Delta T = +\sigma T\alpha/C_v$, so it will undergo a further thermal strain,

$$e_{\text{therm}} = \alpha\Delta T = \sigma T\alpha^2/C_v. \qquad (7.22)$$

At room temperature, the total strain will be

$$e_{\text{tot}} = \sigma(1/E_a + T\alpha^2/C_v). \qquad (7.23)$$

This must be the same as the strain that would have resulted from stressing the material isothermally (so slowly that it remained at room temperature), so $e_{\text{iso}} = \sigma/E_i$, where E_i is the isothermal modulus. Equating $e_{\text{iso}} = e_{\text{tot}}$, $1/E_i = 1/E_a + T\alpha^2/C_p$, or

$$\Delta E/E = (E_a - E_i)/E_a = E_i T\alpha^2/C_p. \qquad (7.24)$$

EXAMPLE PROBLEM #7.3: A piece of metal is subjected to a cyclic stress of 80 MPa. If the phase angle $\delta = 0.1°$ and no heat is transferred to the surroundings, what will be the temperature rise after 100,000 cycles? Data: $E = 80$ GPa, $C = 800$ J/kg-K and $\rho = 4.2$ Mg/m^3.

Solution: $U = (1/2)\sigma^2/E$. $\Delta U/U = 2\pi\sin\delta.\Delta U = 2\pi\sin\delta[(1/2)\sigma^2/E] = \sin(0.1°)(80 \times 10^6$ J/m$^3)^2/(80 \times 10^9$ J/m$^3) = 219$ J/m^3 per cycle. The heat released in 100,000 cycles would be 2.19×10^7 J/m^3. This would cause a temperature increase of $Q/(\rho C) = 2.19 \times 10^7$ J/m^3/[(4200 kg/m^3)(800 J/kg-K)] $= 6.5°$C.

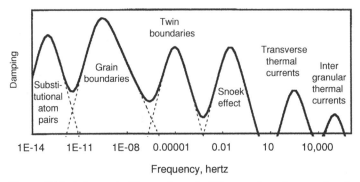

Figure 7.16. Schematic illustration of different damping mechanisms in metals as a function of frequency.

The thermoelastic effect causes damping. The frequencies of the peaks depend on the time for thermal diffusion. The relaxation time, τ, is given by

$$\tau \approx x^2/D, \qquad (7.25)$$

where x is the thermal diffusion distance and D is the thermal diffusivity. The relaxation time, τ, and therefore the frequencies of the damping peaks depend strongly on the diffusion distance, x. There are thermal currents in a polycrystal between grains of different orientations because of their different elastic responses. In this case the diffusion distance, x, is comparable to the grain size, d. For specimens in bending, there are also thermal currents from one side of the specimen to the other. These cause damping peaks with a diffusion distance, x, comparable to the specimen size.

Other Damping Mechanisms

Many other mechanisms contribute to damping in metals. These include the Snoek effect, which involves carbon atoms jumping from one interstitial site to another, the bowing and unbowing of pinned dislocations, increased and decreased dislocation density in pile-ups, viscous grain-boundary sliding, opening and closing of microcracks, and possibly movement of twin boundaries and magnetic domain boundaries. Figure 7.16 shows very schematically the frequency range of different mechanisms in metals.

In thermoplastic polymers, the largest damping peak (α) is associated with the glass transition temperature, which is sensitive to the flexibility of the main backbone chain. Bulky side groups increase the stiffness of the backbone, which increases the glass transition temperature. With many polymers there are more than one damping peak, as indicated in Figure 7.17.

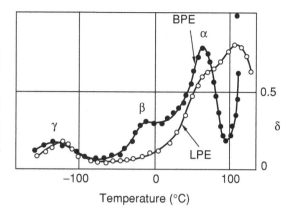

Figure 7.17. Damping peaks in linear polyethylene, LPE, and branched polyethylene, BPE. The α-peak is associated with motions of the backbone, and the β-peak with branches. Linear PE has no branches and hence no b-peak. From N. G. McCrum, C. P. Buckley and C. B. Bucknell, *Principles of Polymer Engineering,* Oxford Science Pub. (1988).

Notes

Rubber is unlike most other materials. On elastic stretching, it undergoes an adiabatic heating, $\Delta T = -\sigma \alpha T / C_v$, because under tension α is negative. If one stretches a rubber band and holds it long enough for its temperature to come to equilibrium with the surroundings and then releases it, the cooling on contraction can be sensed by holding it to the lips. The negative of rubber under tension is the basis for an interesting demonstration. A wheel can be built with spokes of stretched rubber bands extending form hub to rim. When a heat lamp is placed so that it heats half of the spokes, those spokes will shorten, unbalancing the wheel so that it rotates. The direction of rotation is just opposite of the direction that a wheel with metal spokes would rotate.

Bells should have a very low damping. They are traditionally made from bronze that has been heat-treated to form a microstructure consisting of hard intermetallic compounds.

One of the useful characteristics of gray cast iron is the large damping capacity that results from the flake structure of graphite. Acoustic vibrations are absorbed by the graphite flakes. The widespread use of gray cast iron for bases of tool machines (e.g., beds of lathes, drill presses, and milling machines) makes use of this damping capacity.

Coulomb (1736–1806) invented a machine for measuring damping with a torsion pendulum (Figure 7.18).

REFERENCES

C. M. Zener, *Elasticity and Anelasticity of Metals*, U. of Chicago Press, Chicago (1948).

N. G. McCrum, C. P. Buckley and C. B. Bucknell, *Principles of Polymer Engineering*, Oxford Science Pub. (1988).

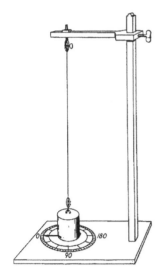

Figure 7.18. Coulomb's torsion pendulum for measuring damping. It is very similar in principle to modern torsion pendulums used for damping studies today. From S. Timoshenko, *History of the Strength of Materials,* McGraw-Hill (1953).

Problems

1. Consider a viscoelastic material whose behavior is adequately described by the combined series-parallel model. Let it be subjected to the force vs. time history shown in Figure 7.19. There is a period of tension followed by compression and tension again. After that, the stress is 0. Let the time interval $\Delta t = K_v / K_{e2} = \tau_e$, and assume $K_{e1} = K_{e2}$. Sketch as carefully as possible the corresponding variation of strain with time.

2. Consider a viscoelastic material whose behavior is adequately described by the combined series-parallel model. Let it be subjected to the strain vs. time history shown in Figure 7.20. There is a period of tension followed by compression and tension again. After that, the stress is returns to zero. Let the time interval $\Delta t = K_v / K_{e2} = \tau_e$, and assume $K_{e1} = K_{e2}$. Sketch as carefully as possible the corresponding variation of force with time.

3. An elastomer was suddenly stretched in tension and the elongation was held constant. After 10 minutes, the tensile stress in the polymer dropped

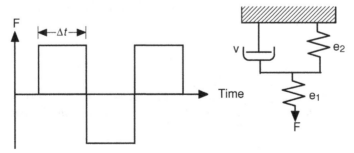

Figure 7.19. Loading of a viscoelastic material.

Table 7.1. *Data for aluminum*

Misc. data for aluminum:			
crystal structure	fcc	lattice parameter	0.4050 nm
density:	2.70 Mg/m^3	Young's modulus	62 GPa
heat capacity	900 J/kg·K	melting point	660 °C

Table 7.2. *Amplitudes of vibration*

Cycle	Amplitude
0	250
25	206
50	170
100	115
200	53

by 12%. After an extremely long time, the stress dropped to 48% of its original value.

A. Find the relaxation time, τ, for stress relaxation.

B. How long will it take the stress to drop to 75% of its initial value?

[Hint: ΔU can be found from the temperature rise, and U can be found from the applied stress and Young's modulus.]

4. A certain bronze bell is tuned to middle C (256 Hz). It is noted that the intensity of the sound drops by one decibel (i.e., 20.56%) every 5 seconds. What is the phase angle δ in degrees?

5. A piece of aluminum is subjected to a cyclic stress of ±120 MPa. After 5000 cycles, it is noted that the temperature of the aluminum has risen by 1.8°C. Calculate Λ and the phase angle δ, assuming that there has been no transfer of heat to the surroundings and all the energy loss/cycle is converted to heat. Data for aluminum are given in Table 7.1.

6. Measurements of the amplitude of vibration of a freely vibrating beam are given in Table 7.2.

A. Calculate the log decrement, Λ.

B. Is Λ dependent on the amplitude for this material in the amplitude range studied? (Justify your answer.)

Figure 7.20. Strain cycles imposed on a vis-coelastic material.

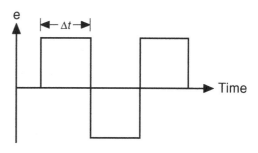

C. What is the phase angle δ?

[Hint: Assume Λ is constant so $e_n/e_{n+1} = e_{n+1}/e_{n+2} =$ and so on. Then $e_n/e_{n+m} = [e_n/e_{n+1}]^m$.]

7 A. A high-strength steel can be loaded up to 100,000 psi in tension before any plastic deformation occurs. What is the largest amount of thermoelastic cooling that can be observed in this steel at 20°C?

B. Find the ratio of adiabatic Young's modulus to isothermal Young's modulus for this steel at 20°C.

C. A piece of this steel is adiabatically strained elastically to 10^{-3} and then allowed to reach thermal equilibrium with the surroundings (20°C) at constant stress. It is then unloaded adiabatically, and again allowed to reach thermal equilibrium with its surroundings. What fraction of the initial mechanical energy is lost in this cycle? [Hint: Sketch the $\sigma-\varepsilon$ path.] For iron, $\alpha = 11.76 \times 10^{-6}/°C$, $E = 29 \times 10^6$ psi, $C_v = 0.46$ J/g°C, and $\rho = 7.1$ g/cm^3.

8. For iron, the adiabatic Young's modulus is $(1 + 2.3 \times 10^{-3})$ multiplied by the isothermal modulus at room temperature. If the anelastic behavior of iron is modeled by a series-parallel model, what is the ratio of K_1 to K_2?

9. Damping experiments on iron were made using a torsion pendulum with a natural frequency of 0.65 cycles/sec. The experiments were run at various temperatures and the maximum log decrement was found at 35°C. The activation energy for diffusion of carbon in α-iron is 78.5 kJ/mol. At what temperature would you expect the damping peak to occur if the pendulum were redesigned so that it had a natural frequency of 10 Hertz?

10. A polymer is subjected to a cyclic stress of 5 MPa at a frequency of 1 Hz for 1 minute. The phase angle is 0.05°. Calculate the temperature rise assuming no loss to the surroundings. $E = 2$ GPa, $C = 1.0$ J/kg $\cdot$ K and $\rho = 1.0$ Mg/m^3.

8 Creep and Stress Rupture

Introduction

Creep is time-dependent plastic deformation that is usually significant only at high temperatures. Figure 8.1 illustrates typical creep behavior. As soon as the load is applied, there is an instantaneous elastic response, followed by period of transient creep (Stage I). Initially the rate is high, but it gradually decreases to a steady state (Stage II). Finally the strain rate may increase again (Stage III), accelerating until failure occurs.

Creep rates increase with higher stresses and temperatures. With lower stresses and temperatures, creep rates decrease but failure usually occurs at lower overall strains (Figure 8.2).

The acceleration of the creep rate in Stage III occurs because the true stress increases during the test. Most creep tests are conducted under constant load (constant engineering stress). As creep proceeds, the cross-sectional area decreases so the true stress increases. Porosity develops in the later stages of creep, further decreasing the load-bearing cross section.

Creep Mechanisms

Viscous flow: Several mechanisms may contribute to creep. These include viscous flow, diffusional flow, and dislocation movement. Viscous flow is the dominant mechanism in amorphous materials

In polycrystalline materials, grain-boundary sliding is viscous in nature. The sliding velocity on the boundary is proportional to the stress and inversely proportional to the viscosity, η. The rate of extension depends on the amount of grain boundary area per volume and is therefore inversely proportional to the grain size, d, so $\dot{\varepsilon} = C(\sigma/\eta)/d$. Viscous flow is thermally activated, so $\eta = \eta_o\exp(Q_V/RT]$. The strain-rate attributable to grain-boundary sliding can be written as

$$\dot{\varepsilon}_V = A_V(\sigma/d)\exp(-Q_V/RT]. \tag{8.1}$$

117

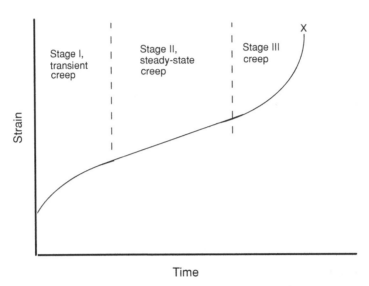

Figure 8.1. Typical creep curve showing three stages of creep.

If grain boundary sliding were the only active mechanism, there would be an accumulation of material at one end of each boundary on which sliding occurs and a deficit at the other end, as sketched in Figure 8.3. This incompatibility must be relieved by another deformation mechanism, one involving dislocation motion or diffusion.

Diffusion-controlled creep: A tensile stress increases the separation of atoms on grain boundaries that are normal to the stress axis, and the Poisson contraction decreases the separation of atoms on grain boundaries that are parallel to the stress axis. The result is a driving force for the diffusional transport of atoms from grain boundaries parallel to the tensile stress to boundaries normal to the tensile stress. Such diffusion produces a plastic elongation, as shown in Figure 8.4. The specimen elongates as atoms are added to grain boundaries perpendicular to the stress.

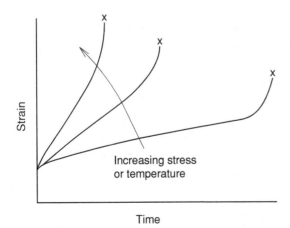

Figure 8.2. Decreasing temperatures and stresses lead to slower creep rates, with failures often occurring at a lower strains.

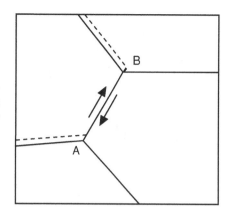

Figure 8.3. Grain boundary sliding causes incompatibilities at both ends of the planes, *A* and *B*, on which sliding occurs. This must be relieved by another mechanism for sliding to continue.

If the creep occurs by diffusion through the lattice, it is called *Nabarro-Herring creep*. The diffusional flux, J, between the boundaries parallel and perpendicular to the stress axis is proportional to the stress, σ, and the lattice diffusivity, D_L, and is inversely proportional to the diffusion distance, $d/2$, between the diffusion source and sink. Therefore $J = C D_L \sigma / (d/2)$, where C is a constant. The velocity, v, at which the diffusion source and sink move apart is proportional to the diffusion flux, so $v = C D_L \sigma / (d/2)$. Because the strain rate equals $v/(d/2)$,

$$\dot{\varepsilon}_{N-H} = A_L (\sigma/d^2) D_L \qquad (8.2)$$

where A_L is a constant.

On the other hand, if creep occurs by diffusion along the grain boundaries it is called *Coble creep*. The driving force for Coble creep is the same as for Nabarro-Herring creep. The total number of grain-boundary diffusion paths is inversely proportional to the grain size, so now J is proportional to $d^{-1/3}$ and

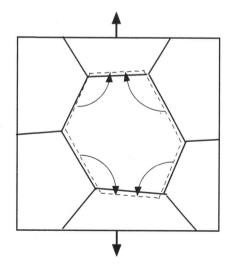

Figure 8.4. Creep by diffusion between grain boundaries. As atoms diffuse from lateral boundaries to boundaries normal to the tensile stress, the grain elongates and contracts laterally.

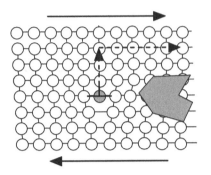

Figure 8.5. Climb-controlled creep. An edge dislocation can climb by diffusion of atoms away from the dislocation (vacancies to the dislocation), thereby avoiding an obstacle.

the creep rate is given by

$$\dot{\varepsilon}_C = A_G(\sigma/d^3)D_{gb}, \tag{8.3}$$

where D_{gb} is the diffusivity along grain boundaries and A_G is a constant.

Dislocation motion: Slip is another mechanism of creep. In this case, the creep rate is controlled by how rapidly the dislocations can overcome obstacles that obstruct their motion. At high temperatures, the predominant mechanism for overcoming obstacles is dislocation climb (Figure 8.5). With climb, the creep rate is not dependent on grain size but the rate of climb does depend very strongly on the stress,

$$\dot{\varepsilon} = A_s\sigma^m. \tag{8.4}$$

The value of m is approximately 5 for climb-controlled creep.* As climb depends on diffusion, the constant A_s has the same temperature dependence as lattice diffusion. At lower temperatures, creep is not entirely climb-controlled and larger exponents are observed.

Equations 8.1 through 8.4 predict creep rates that depend only on stress and temperature and not on strain. Thus, they apply only to Stage II or steady-state creep.

Multiple mechanisms: More than one creep mechanism may be operating. There are two possibilities; either the mechanisms operate independently or they act cooperatively. If they operate independently, the overall creep rate, $'\varepsilon$, is the sum of the rates due to each mechanism,

$$\dot{\varepsilon} = \dot{\varepsilon}_A + \dot{\varepsilon}_B + \cdots \tag{8.5}$$

on the most rapid mechanism.

On the other hand, two mechanisms may be required to operate simultaneously, as in the case grain-boundary sliding requiring another mechanism.

---

* This notation widely used in the creep literature is the inverse of that used in Chapter 6, where we wrote $\sigma \propto \dot{\varepsilon}^m$ or, equivalently, $\dot{\varepsilon} \propto \dot{\varepsilon}^{1/m}$. The m used in the creep literature is the reciprocal of the m used earlier. The value of m = 5 here corresponds to a value of $m = 0.2$ in the notation of Chapter 6.

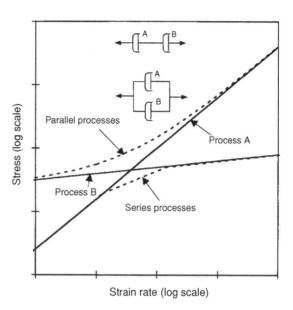

Figure 8.6. Creep by two mechanisms, A and B. If the mechanisms operate independently (series), the overall creep rate is largely determined by the faster mechanism. If creep depends on the operation of both mechanisms (parallel), the potentially slower mechanism will control the overall creep rate.

Where two parallel mechanisms are required, both must operate at the same rate,

$$\dot{\varepsilon} = \dot{\varepsilon}_A = \dot{\varepsilon}_B. \qquad (8.6)$$

The overall rate is determined by the potentially slower mechanism. These two possibilities (Equations 8.5 and 8.6) are illustrated in Figure 8.6.

Cavitation

Cavitation can lead to fracture during creep. Cavitation occurs by nucleation and growth of voids, particularly at grain boundaries and second-phase particles. A void will grow if its growth lowers the energy of the system. Consider the growth of a spherical void in a cubic element with dimensions x, as illustrated in Figure 8.7. Let the radius be r and the stress σ. The surface energy of the void is $E_S = 4\pi\gamma r^2$. If the radius increases by dr, the increase of the surface energy is

$$dU_s = 8\pi\gamma r\,dr. \qquad (8.7)$$

The volume of the void is $(4/3)\pi r^3$. Growth by dr will change the volume of the sphere by $4\pi r^2 dr$. If all of the atoms diffuse to positions that cause lengthening, the element will lengthen by dx so that its volume increase is $x^2 dx = 4\pi r^2 dr$. The strain, de, associated with the growth will be $de = dx/x = 4\pi r^2 dr/x^3$. The energy per volume expended by the applied stress, σ, is $dU_s = \sigma\,de = 4\sigma\pi r^2 dr/x^3$. The total energy associated with the element is

$$dU_s = x^3 4\sigma\pi r^2 dr/x^3 = 4\sigma\pi r^2 dr. \qquad (8.8)$$

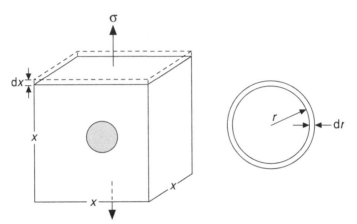

Figure 8.7. Spherical void in a cubic cell. Growth of the void by dr causes an elongation of the cell by dx.

Equating 8.7 and 8.8, the critical condition for void growth is

$$r^* = 2\gamma/\sigma. \tag{8.9}$$

Voids with radii less than r^* should shrink. Ones larger than r^* should grow.

Failure may also occur by grain boundary fracture. Unless grain boundary sliding is accompanied by another mechanism, grain boundary cracks must form as illustrated in Figure 8.8.

Rupture vs. Creep

The term *rupture* is used in the creep literature to mean fracture. Up to this point, the discussion has centered on creep deformation. Creep may cause component parts to fail in service by excessive deformation, so they no longer can function satisfactorily. However, after long times under load at high temperature, it is more common that parts fail by rupture than by excessive deformation. As the service temperatures and stress level are lowered to achieve lower rates of creep, rupture usually occurs at lower strains. Figure 8.9 shows this for a Ni-Cr-Co-Fe alloy. Successful design for high-temperature service must ensure against both creep excessive deformation and against fracture (creep-rupture).

Figure 8.8. Schematic views of grain boundary cracks caused by grain boundary sliding.

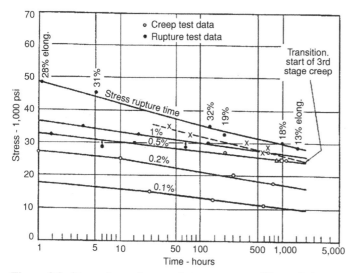

Figure 8.9. Stress dependence of stress-rupture life and time to reach several creep strains for a Ni-Cr-Co-Fe alloy tested at 650°C. Note that as the stress level is reduced to increase the time to a given strain, rupture occurs at lower strains. From N. J. Grant in *High Temperature Properties of Metals*, ASM (1950).

Extrapolation Schemes

For many applications such as power-generating turbines, boilers, engines for commercial jets, and furnace elements, components must be designed for long service at high temperatures. Some parts are designed to last 20 years (175,000 hrs) or more. In such cases, it is not feasible to test alloys under creep conditions for times as long as the design life. Often tests are limited to 1000 hrs (42 days). To ensure that neither rupture nor excessive creep strain occurs, results from shorter time tests at higher temperatures and/or higher stresses must be extrapolated to the service conditions. Several schemes have been proposed for such extrapolation.

Sherby-Dorn: Sherby and Dorn proposed plotting creep strain as a function of a temperature-compensated time,

$$\theta = t \exp(-Q/\mathrm{R}T). \qquad (8.10)$$

In creep studies, $\log_{10}\theta$ is often referred to as the Sherby-Dorn parameter, P_{SD}. Note that θ is the same as the Zener-Hollomon parameter discussed in Chapter 6. The basic idea is that the creep strain is plotted against the value of θ for a given stress, as shown in Figure 8.10. For a given material and stress, the creep strain depends only on θ. Alternatively, the stress to rupture or to achieve a given amount of creep strain is plotted against the value of P_{SD}, as shown in Figure 8.11. From such plots, one can predict long-time behavior.

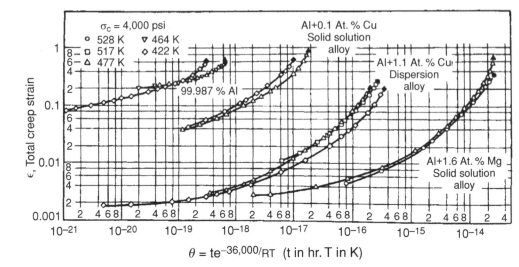

$$\theta = te^{-36,000/RT} \quad (t \text{ in hr. } T \text{ in } K)$$

Figure 8.10. Creep strain vs. θ for several aluminum alloys tested at 27.6 MPa. Note that the test data for several temperatures fall nearly on the same line. From R. L. Orr, O. D. Sherby and J. E. Dorn, *Trans ASM*, v. 96 (1954).

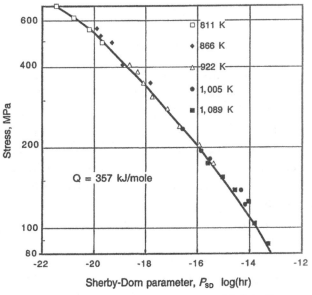

Sherby-Dorn parameter, P_{SD} log(hr)

Figure 8.11. Stress to cause creep rupture of S-590 alloy as a function of PSD = $\log_{10}\theta$. The value of Q was taken as 357 J/mol. Note that the test data for several temperatures fall on the same line. Data taken from R. M. Goldoff, *Materials in Engineering Design*, v. 49 (1961).

EXAMPLE #8.2: The chief engineer needs to know the permissible stress on the alloy in Figure 8.13 to achieve a rupture life of 20 years at 500°C.

Solution: For these conditions, $\theta = (20 \times 365 \times 24 \text{ hrs})\exp\{-85,000/[1.987(500 + 273)]\} = 1.62 \times 10^{-19}$, so $\log_{10}(1.62 \times 10^{-19}) = -18.8$. Figure 8.11 indicates that the permissible stress would be about 400 MPa.

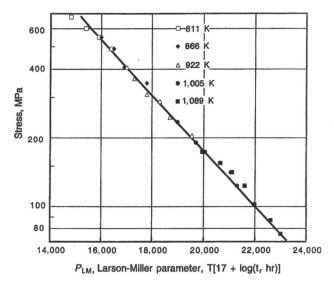

Figure 8.12. Stress to cause creep rupture of alloy S-590 as a function of the Larson-Miller parameter P. This figure was constructed with the same data as used in Figure 8.13 with $C = 17$ and temperature in Kelvin. Note that the test data for several temperatures fall on the same line.

Larson-Miller parameter: A different extrapolation parameter was proposed by Larson and Miller,[*]

$$P_{LM} = T(\log_{10}t_r + C). \tag{8.11}$$

As they originally formulated this parameter, the temperature, T, was expressed as $T + 460$, where T is in degrees Fahrenheit and the rupture time, t_r, in hours. They found agreement for most high-temperature alloys with values of C of about 20. However, the parameter can be expressed with temperature in Kelvin, keeping t_r in hours. Figure 8.12 is a Larson-Miller plot of stress for rupture vs. P. Again, one can use such plots for prediction of long-term behavior from shorter time tests at higher temperatures.

> **EXAMPLE #8.3:** The chief engineer needs to know the permissible stress that can be applied to the S-590 alloy for a life of 20 years at 500°C.
>
> *Solution:* The Larson-Miller parameter with C = 17 is $P = 773[17 + \log_{10}$ $(20 \times 365 \times 24\,\text{hrs})] = 17{,}190$. From Figure 8.14, $P = 17{,}190$ corresponds to $\sigma \approx 400$ MPa.

REFERENCES

N. E. Dowling, *Mechanical Behavior of Materials*, 2nd Ed, Prentice-Hall (1999).

M. A. Meyers and K. K. Chawla, *Mechanical Behavior of Materials*, Prentice-Hall (1999).

* F. R. Larson and J. Miller, *Trans ASME,* v. 74 (1952), pp. 765–771.

Figure 8.13. Andrade's constant true-stress creep apparatus. As the specimen elongates, the buoyancy of the water lowers the force on the specimen.

T. H. Courtney, *Mechanical Behavior of Materials*, 2nd Ed, McGraw-Hill, 2000.

R. W. Hertzberg, *Deformation and Fracture Mechanics of Engineering Materials*, 4th Ed, John Wiley (1995).

Notes

In the paper that proposed their extrapolation scheme, Larson and Miller* rationalized the value of about 20 for the constant, C, in their parameter. They noted that if T were infinite, P would be infinite unless $\log_{10}t_r + C = 0$, so $C = -\log_{10}t_r$. Furthermore, they assumed that at an infinite temperature the pieces of a breaking specimen should fly apart with the speed of light, c. Taking the fracture to occur when the pieces are separated by an atomic diameter, d, the time, t, for fracture is d/c. Assuming $d = 0.25$ nm as a typical value, $t = (0.25 \times 10^{-9}\text{m}/3 \times 10^8\text{m/s})(3600\text{s/hr}) = 2.3 \times 10^{-22}$hr. $C = -\log_{10}t_r = 21.6$, or approximately 20.

In 1910, E. M. daC. Andrade* proposed a scheme for maintaining constant true stress during a creep experiment by using dead-weight loading with a shaped weight partially suspended in water (Figure 8.13). As the specimen elongates, more of the weight is immersed, decreasing the force on the specimen.

Problems

1. Stress vs. rupture life data for a super alloy are listed in Table 8.1. The stresses are given in MPa and rupture life is given in hours.

   A. Make a plot of stress (log scale) vs. P_{LM} where $P_{LM} = (T)(C + \log_{10}t)$. T is the temperature in Kelvin, t is in hours, and $C = 20$.

---

* E. M. daC. Andrade, *Proc.Roy. Soc. A London*, v. 84 (1910), pp. 1–12.

Table 8.1. *Stress-rupture data*

Stress MPa	rupture time (hrs) 500°C	600°C	700°C	800°C
600	2.8	0.018	0.0005	–
500	250.	0.72	0.004	–
400	–	12.1	0.082	0.00205
300	–	180	0.87	0.011
200	–	2412	11.0	0.198
100	–	–	98.0	1.10

B. Predict from that plot what stress would cause rupture in 100,000 hrs at 450°C.

2. A. Using the data in Problem 1, plot the Sherby-Dorn parameter, $P_{SD} = \log\theta$ where $\theta = t\exp(-Q/RT)$ and $Q = 340\,\text{kJ/mole}$.

   B. Using this plot, predict what stress would cause rupture in 100,000 hrs at 450°C.

3. For many materials, the constant C in the Larson-Miller parameter, $P = (T + 460)(C + \log_{10}t)$ (where T is in Fahrenheit, and t in hours), is equal to 20. However, the Larson-Miller parameter can also be expressed as $P' = T(C' + \ln t)$ with t in seconds and T in Kelvin, using the natural logarithm of time. In these cases, what is the value of C'?

4. Stress rupture data is sometimes correlated with the Dorn parameter, $\theta = t\exp[-Q/(RT)]$, where t is the rupture time, T is absolute temperature, and θ is assumed to depend only on stress. If this parameter correctly describes a set of data, then a plot of $\log(t)$ vs. $1/T$ for data at a single level

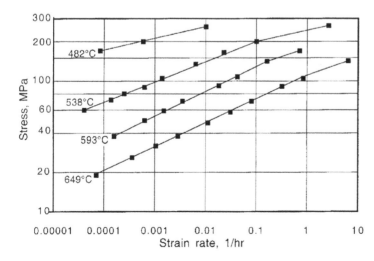

Figure 8.14. Creep data for a carbon steel. Data from P. N. Randall, *Proc, ASTM*, v. 57 (1957).

Table 8.2. *Creep for stress data*

at 650°C	stress (ksi)	80	65	60	40
	rupture life (hrs)	0.08	8.5	28	483
at 730°C	stress (ksi)	60	50	30	25
	rupture life (hrs)	0.20	1.8	127	1023
at 815°C	stress (ksi	50	30	20	
	rupture life (hrs)	0.30	3.1	332	
at 925°C	stress (ksi)	30	20	15	10
	rupture life (hrs)	0.08	1.3	71	123
at 1040°C	stress (ksi)	20	10	5	
	rupture life (hrs)	0.03	1.0	28	211

of stress would be a straight line. If the Larson-Miller parameter correctly correlates data, a plot of data at constant stress (therefore constant P) of log(t) vs. $1/T$ also would be a straight line.

A. If both parameters predict straight lines on $\log(t)$ vs. $1/T$ plots, are they really the same thing?

B. If not, how do they differ? How could you tell from a plot of $\log(t)$ vs. $(1/T)$ which parameter better correlates a set of stress rupture data?

5. Data for the steady-state creep of a carbon steel are plotted in Figure 8.14.

A. Using the linear portions of the plot, determine the exponent m in $\dot{\varepsilon}_{sc} = B\sigma^m$ at 538°C and 649°C. Determine the activation energy, Q, in the equation $\dot{\varepsilon} = f(\sigma)\exp[-Q/(RT)]$.

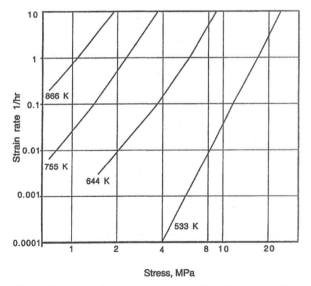

Figure 8.15. Steady state creep rate of an aluminum alloy at several temperatures. Data from O. D. Sherby and P. M. Burke, in *Prog. Mater. Sci.* v. 13 (1968).

6. The data in Table 8.2 were obtained in a series of stress rupture tests on a material being considered for high temperature service.
   A. Make a Larson-Miller plot of the data.
   B. Predict the life for an applied stress of 30 ksi at 600°C.
7. Consider the creep rate vs. stress curves for an aluminum alloy plotted in Figure 8.15. Calculate the stress exponent, m, at 755 K.

9 Ductility and Fracture

Introduction

Throughout history, there has been a neverending effort to develop materials with greater yield strengths. However, a greater yield strength is generally accompanied by a lower ductility and a lower toughness. *Toughness* is the energy absorbed in fracturing. A high-strength material has low toughness because it can be subjected to greater stresses. The stress necessary to cause fracture may be reached before there has been much plastic deformation to absorb energy. Ductility and toughness are lowered by factors that inhibit plastic flow. As schematically indicated in Figure 9.1, these factors include decreased temperatures, increased strain rates, and the presence of notches. Developments that increase yield strength usually result in lower toughness.

In many ways, the fracture behavior of steel is like that of taffy candy. It is difficult to break a warm bar of taffy candy to share with a friend. Even children know that warm taffy tends to bend rather than break. However, there are three ways to promote its fracture. A knife may be used to notch the candy bar, producing a stress concentration. The candy may be refrigerated to raise its resistance to deformation. Finally, rapping it against a hard surface raises the loading rate, increasing the likelihood of fracture. Notches, low temperatures, and high rates of loading also embrittle steel.

There are two important reasons for engineers to be interested in ductility and fracture. The first is that a reasonable amount of ductility is required to form metals into useful parts by forging, rolling, extrusion, or other plastic working processes. The second is that a certain degree of toughness is required to prevent failure in service. Some plastic deformation is necessary to absorb energy.

This chapter treats the mechanisms and general observations of failure under a single application of load at low and moderate temperatures. Chapter 8 covers failure under creep conditions at high temperatures, and fatigue failure under cyclic loading is covered in Chapter 11.

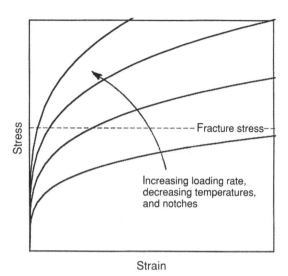

Figure 9.1. Lowered temperatures, increased loading rates, and the presence of notches all reduce ductility. These three factors raise the stress level required for plastic flow so the stress required for fracture is reached at lower strains.

Fractures can be classified several ways. A fracture is described as *ductile* or *brittle* depending on the amount of deformation that precedes it. Failures may also be described as *intergranular* or *transgranular*, depending on the fracture path. The terms *cleavage, shear, void coalescence*, and so on are used to identify failure mechanisms. These descriptions are not mutually exclusive. A brittle fracture may be intergranular or it may occur by cleavage.

The *ductility* of a material describes the amount of deformation that precedes fracture. Ductility may be expressed as percent elongation or as the percent reduction of area in a tension test. Failures in tension tests may be classified several ways (Figure 9.2). At one extreme, a material may fail by necking down to a vanishing cross-section. At the other extreme, fracture may occur

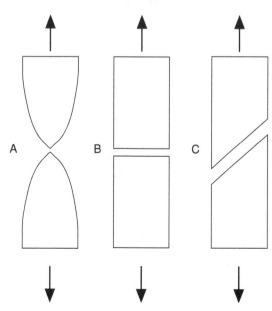

Figure 9.2. Several failure modes. (A) Rupture by necking down to a zero cross section. (B) Fracture on a surface that is normal to the tensile axis. (C) Shear fracture.

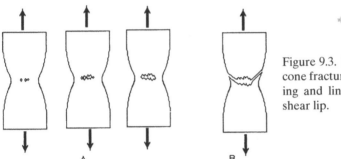

Figure 9.3. Development of a cup and cone fracture. (A) Internal porosity growing and linking up. (B) Formation of a shear lip.

on a surface that is more-or-less normal to the maximum tensile stress with little or no deformation. Failures may also occur by shear.

Ductile Fracture

Failure in a tensile test of a ductile material occurs well after the maximum load is reached and a neck has formed. In this case, fracture usually starts by nucleation of voids at inclusions in the center of the neck where the hydrostatic tension is the greatest. As deformation continues, these internal voids grow and eventually link up by necking of the ligaments between them (Figures 9.3, 9.4, and 9.5). Such a fracture starts in the center of the bar, where the hydrostatic tension is greatest. With continued elongation, this internal fracture grows outward until the outer rim can no longer support the load and the edges fail by sudden shear. The final shear failure at the outside also occurs by void formation and growth (Figures 9.6 and 9.7). This overall failure is often called a *cup and cone* fracture. If the entire shear lip is on the same broken piece, it forms a cup. The other piece is the cone (Figure 9.8). More often, part of the shear lip is on one half of the specimen and part on the other half.

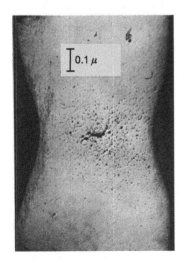

Figure 9.4. Section through a necked tensile specimen of copper, showing an internal crack formed by linking voids. From K. E. Puttick, *Phil Mag.* v. 4 (1959).

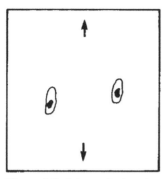 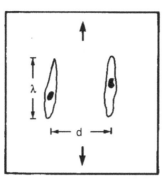 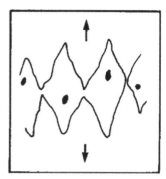

Figure 9.5. Schematic drawing showing the formation and growth of voids during tension, and their linking up by necking of the ligaments between them. From W. F. Hosford and R. M. Caddell, *Metal Forming: Mechanics and Metallurgy, 3rd Ed.*, Cambridge (2007).

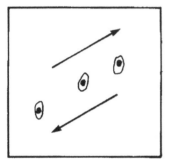 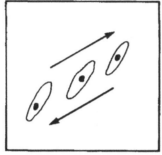 

Figure 9.6. Schematic drawing illustrating the formation of and growth of voids during shear, their growth, and linking up by necking of the ligaments between them. From Hosford and Caddell, *ibid*.

Figure 9.7. Large voids in a localized shear band in OFHC copper. From H. C. Rogers, *Trans. TMS-AIME* v. 218 (1960).

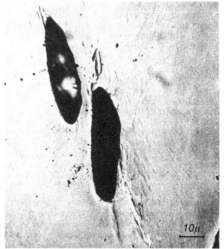

Figure 9.8. A typical cup and cone fracture in a tension test of a ductile manganese bronze. From *Elements of Physical Metallurgy*, A. Guy, Addison-Wesley (1959).

In ductile materials voids form at inclusions because either the inclusion-matrix interface or the inclusion itself is weak. Figure 9.9 shows the fracture surface formed by coalescence of voids. The inclusions can be seen in some of the voids. Ductility is strongly dependent on the inclusion content of the material. With increasing numbers of inclusions, the distance between the voids decreases, so it is easier for them to link together and lower the ductility. Figure 9.10 shows the decrease of ductility of copper with volume fraction inclusions. Ductile fracture by void coalescence can occur in shear as well as in tension testing.

The level of hydrostatic stress plays a dominant role in determining the fracture strains. Hydrostatic tension promotes the formation and growth of voids, whereas hydrostatic compression tends to suppress void formation and growth. Photographs of specimens tested in tension under hydrostatic pressure (Figure 9.11) show that the reduction of area increases with pressure. Figure 9.12 shows how the level of hydrostatic stress affects ductility.

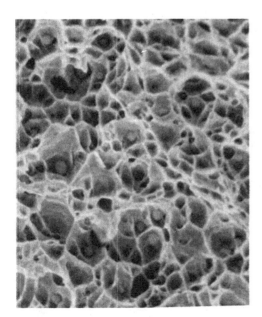

Figure 9.9. Dimpled ductile fracture surface in steel. Note the inclusions associated with about half the dimples. The rest of the inclusions are on the mating surface. Courtesy of J. W. Jones. From Hosford and Caddell, *ibid*.

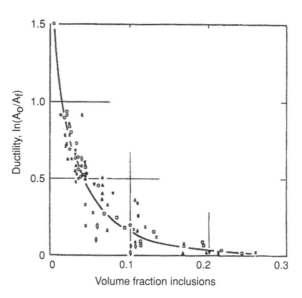

Figure 9.10. Effect of second-phase particles on the tensile ductility of copper. Data include alumina, silica, molybdenum, chromium, iron, and iron-molybdenum inclusions as well as holes. From B. I. Edelson and W. M. Baldwin *Trans. Q. ASM.* v. 55 (1962).

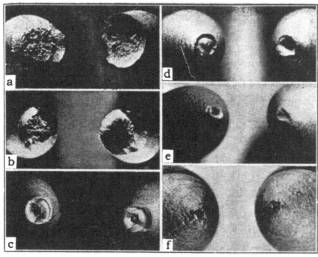

Figure 9.11. Effect of pressure on the area reduction in tension tests of steel. (a) atmospheric pressure, (b) 234 kPa, (c) 1 MPa, (d) 1.3 MPa, (e) 185 MPa, and (f) 267 MPa. From P. W. Bridgman in *Fracture of Metals,* ASTM (1947).

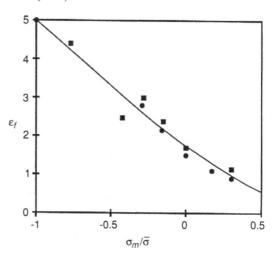

Figure 9.12. The effective fracture strain, $\bar{\varepsilon}_f$, increases as the ratio of the mean stress, σ_m, to the effective stress, $\bar{\sigma}$, becomes more negative (compressive) for two of the steels tested by Bridgman. Data from A. L. Hoffmanner, Interim report, Air Force contract 33615–67-C (1967).

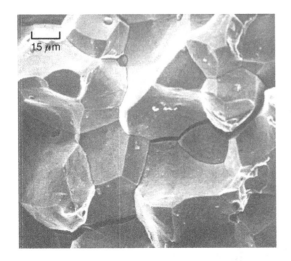

Figure 9.13. Intergranular fracture in pure iron under impact. From *Metals Handbook, 8th ed.*, v. 9, ASM (1974).

Void Failure Criterion

To explain the role of voids in failure, Gurson[*] proposed an upper-bound damage for a porous material that contains voids and obeys the von Mises yield criterion. His yield criterion is of the form

$$F = \left(\frac{\bar{\sigma}}{\sigma_f}\right)^2 + 2f \cosh\left(\frac{3}{2}\frac{\sigma_H}{\sigma_f}\right) - (1 + f^2) = 0, \qquad (9.1)$$

where f is the volume fraction voids, s_o is the yield strength, $\bar{\sigma}$ is the yield strength, and σ_H is the hydrostatic stress. Failure occurs when f is so large that $F = 0$.

Later, Tvergaard[†] suggested that a good fit with experimental data could be obtained by with three fitting parameters c_1, c_2 and c_3:

$$F == \left(\frac{\bar{\sigma}}{\sigma_f}\right)^2 + 2c_1 f \cosh\left(\frac{c_2}{2}\frac{\sigma_H}{\sigma_f}\right) - (1 + c_3 f^2) = 0. \qquad (9.2)$$

Brittle Fracture

In some materials, fracture may occur by *cleavage*. Cleavage fractures occur on certain crystallographic planes (*cleavage planes*) that are characteristic of the crystal structure. It is significant that fcc metals do not undergo cleavage.

Some polycrystals have brittle grain boundaries, which form easy fracture paths. Figure 9.13 shows such an intergranular fracture surface. The brittleness of grain boundaries may be inherent to the material, or may be caused by segregation of impurities to the grain boundary, or even by a film of a brittle second phase. Commercially pure tungsten and molybdenum fail by

[*] A. L. Gurson, *PhD thesis*, Brown University, 1975.
[†] V. Tvergaard, in *Advances in Applied Mechanics*, v 27, (1990) pp. 83–152.

grain boundary fracture. These metals are ductile only when all the grain boundaries are aligned with the direction of elongation, as in tension testing of cold-drawn wire. Copper and copper alloys are severely embrittled by a very small amount of bismuth, which segregates to and wets the grain boundaries. Molten FeS in the grain boundaries of steels at hot working temperatures would cause failure along grain boundaries. Such loss of ductility at high temperatures is called *hot shortness*. Hot shortness is prevented in steels by adding manganese, which reacts with the sulfur to form MnS. Manganese sulfide is not molten at hot-working temperatures and does not wet the grain boundaries. Stress corrosion is responsible for some grain boundary fractures.

With brittle fracture, toughness depends on grain size. Decreasing the grain size increases the toughness and ductility. Perhaps this is because cleavage fractures must re-initiate at each grain boundary, and with smaller grain sizes there are more grain boundaries. Decreasing grain size, unlike most material changes, increases both yield strength and toughness.

Impact Energy

A material is regarded as being *tough* if it absorbs a large amount of energy in breaking. In a tension test, the energy per volume to cause failure is the area under the stress-strain curve and is the toughness in a tension test. However, the toughness under other forms of loading may be very different because toughness depends also on the degree to which deformation localizes. The total energy to cause failure depends on the deforming volume as well as on energy per volume.

Impact tests are often used to assess the toughness of materials. The most common of these is the *Charpy test*. A notched bar is broken by a swinging pendulum. The energy absorbed in the fracture is measured by recording by how high the pendulum swings after the bar breaks. Figure 9.14 gives the details of the test geometry. The standard specimen has a cross-section of 10 mm by 10 mm. There is a 2-mm deep V-notch with a radius of 0.25 mm. The pendulum's mass and height are standardized. Sometimes bars with U- or keyhole notches are employed instead. Occasionally, subsized bars are tested.

One of the principal advantages of the Charpy test is that the toughness can easily be measured over a range of temperatures. A specimen can be heated or cooled to the specified temperature, and then transferred to the Charpy machine and broken quickly enough so that its temperature change is negligible. For many materials, there is a narrow temperature range over which there is a large change of energy absorption and fracture appearance. It is common to define a *transition temperature* in this range. At temperatures below the transition temperature, the fracture is brittle and absorbs little energy in a Charpy test. Above the transition temperature, the fracture is

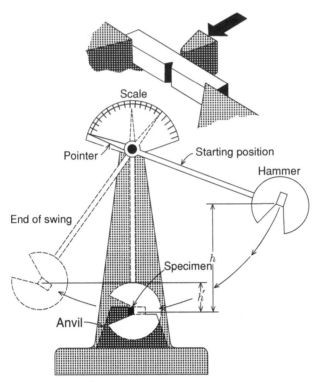

Figure 9.14. Charpy testing machine and test bar. A hammer on the pendulum breaks the bar. The height the pendulum swings after breaking the bar indicates the energy absorbed. From H. W. Hayden, W. G. Moffatt and J. Wulff, *Structure and Properties of Materials, vol. III Mechanical Behavior,* Wiley (1965).

ductile and absorbs a large amount of energy. Figure 9.15 shows typical results for steel.

The transition temperature does not indicate a structural change of the material. The ductile-brittle transition temperature depends greatly on the

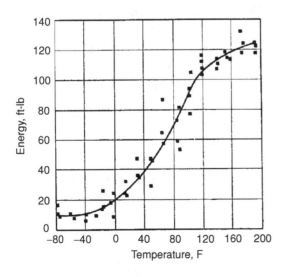

Figure 9.15. Ductile-brittle transition in a Charpy V-notch specimen of a low-carbon, low-alloy hot-rolled steel. From R. W. Vanderbeck and M. Gensamer, *Welding J. Res. Suppl.* (Jan. 1950).

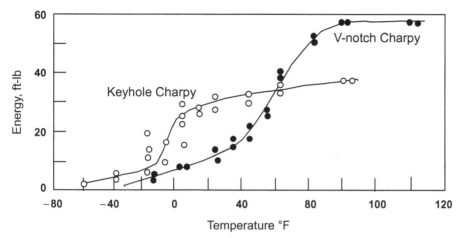

Figure 9.16. Charpy test results for one steel with two different notch geometries (standard V-notch and keyhole notch) depends on the mode of testing. Data from Pellini, *Spec. Tech. Publ.* 158, (1954).

type of test being conducted. With less severe notches, as in keyhole or U-notched Charpy test bars, lower transition temperatures are measured (Figure 9.16). With decreased specimen width, there is less triaxiality so the transition temperatures are lowered. With slow bend tests and unnotched tensile tests, even lower ductile-brittle transitions are observed. In discussing the ductile-brittle transition temperature of a material, one should specify not only the type of test but also the criterion used. Specifications for ship steels area often require a certain *Charpy V-notch 15 ft-lb transition temperature* (the temperature at which the energy absorption in a V-notch Charpy test is 15 ft-lbs).

There are two principal uses for Charpy data. One is in the development of tougher materials. For example, it has been learned through Charpy testing that the transition temperature of hot-rolled carbon steels can be lowered by decreasing the carbon content (Figure 9.17).

The other main use of Charpy data is for documenting and correlating service behavior. This information can then be used in the design and specifications of new structures. For example, a large number of ships failed by brittle fracture during World War II. Tests were conducted on steel from plates in which cracks initiated, plates through which cracks propagated, and a few plates in which cracks stopped. The results were used to establish specifications for the 15 ft-lb Charpy V-notch transition temperature for plates used in various parts of a ship.

In general, bcc metals and many hcp metals exhibit a ductile-brittle transition but it is significant that fcc metals do not. For fcc metals, changes of impact energy with temperature are small, as shown for aluminum in Figure 9.18. Because of this, austenitic stainless steels or copper are frequently used in equipment for cryogenic applications.

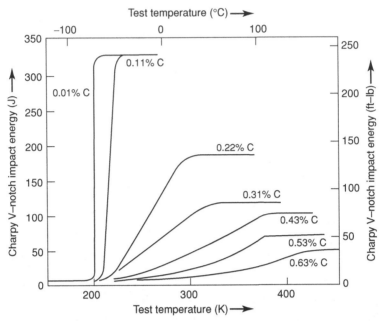

Figure 9.17. Effect of carbon content on Charpy V-notch impact energy. Decreasing carbon content lowers the ductile-brittle transition temperature and raises the shelf energy. From J. A. Rinebilt and W. J. Harris, *Trans. ASM*, v. 43 (1951).

Figure 9.18. A natural gas pipeline that failed during field testing. From E. Parker, *Brittle Fracture of Engineering Structures*, Wiley (1957).

Figure 9.19. A ship that fractured while in port. From C. F. Tipper, *The Brittle Fracture Story,* Cambridge University Press (1963).

Notes

In the period 1948 to 1951, there were many fractures of natural gas pipelines. Most occurred during testing and most started at welding defects but propagated through sound metal. One of the longest cracks was 3200 feet long. Once started, cracks run at speeds greater than the velocity of sound in the pressurized gas. Therefore, there is no release of the gas pressure to reduce the stress at the tip of the crack. Figure 9.18 shows one of the cracked lines.

During World War II, there was a rapid increase of shipbuilding. Production of ships by welding of steel plates together (in contrast to the earlier procedure of joining them by riveting) became common. As a result, a large number of ships, particularly Liberty Ships and T-2 Tankers, failed at sea. More ships sunk as the result of brittle fractures than by German U-boat activity. Recovery of some ships and half-ships allowed the cause of the failures to be investigated. There were three main factors: poor welds, ship design (cracks often started at sharp cornered hatchways that created stress concentrations), and high transition temperatures of the steels. Figure 9.19 is a photograph of a ship that failed in harbor.

Percy W. Bridgman (1882–1961) was born in Cambridge, Massachusetts, and attended Harvard, where he graduated in 1904 and received his Ph.D. in 1908. He discovered the high-pressure forms of ice, and is reputed to have discovered "dry ice" and wrote a classic book, *Physics of High Pressures*, in 1931. In 1946, he was awarded the Nobel prize in physics for his work at high pressures. Another book, *Studies in Large Plastic Flow and Fracture* (1952), summarizes his work with the mechanics of solids.

REFERENCES

Ductility, ASM, Metals Park (1967).
Fracture of Engineering Materials, ASM, Metals Park (1964).
E. Parker, *Brittle Fracture of Engineering Structures*, Wiley (1957).
F. McClintock and A. Argon, *Mechanical Behavior of Materials*, Addison-Wesley (1966).
W. Hosford and R. Caddell, *Metal Forming: Mechanics and Metallurgy* 3rd ed. Cambridge (2007)

Problems

1. Derive the relation between % El and % RA for a material that fractures before it necks. (Assume constant volume and uniform deformation.)

2. Consider a very ductile material that begins to neck in tension at a true strain of 0.20. Necking causes an additional elongation that is approximately equal to the bar diameter. Calculate the percent elongation of this material if the ratio of the gauge length to bar diameter is 2, 4, 10, and 100. Plot percent elongation vs. L_o/D_o.

3. For a material with a tensile yield strength, Y, determine the ratio of the mean stress $\sigma_m =$ to Y at yielding in a
   A. Tension test,
   B. Torsion test, and
   C. Compression test.

4. Cleavage in bcc metals occurs more frequently as the temperature is lowered and as the strain rate is increased. Explain this observation.

5. It has been argued that the growth of internal voids in a material while it is being deformed is given by $dr = f(\sigma_H)d\bar{\varepsilon}$, where r is the radius of the void and $f(\sigma_H)$ is a function of the hydrostatic stress. Using this hypothesis, explain why ductile fracture occurs at greater effective strains in torsion than in tension.

6. Explain why voids often form at or near hard inclusions, both in tension and in compression.

7. Is it safe to say that brittle fracture can be avoided in steel structures if the steel is chosen so that its Charpy V-notch transition temperature is below the service temperature? If not, what is the value of specifying Charpy V-notch test data in engineering design?

8. Hold a piece of newspaper by the upper left corner with one hand and the upper right corner with the other, and tear it. Take another piece of newspaper, rotate it 90° and repeat. One of the tears will be much straighter than the other. Why?

Fracture Mechanics

Introduction

The treatment of fracture in Chapter 9 was descriptive and qualitative. In contrast, *fracture mechanics* provides a quantitative treatment of fracture. It allows measurements of the toughness of materials and provides a basis for predicting the loads that structures can withstand without failure. Fracture mechanics is useful in evaluating materials, in the design of structures, and in failure analysis.

Early calculations of strength for crystals predicted strengths far in excess of those measured experimentally. The development of modern fracture mechanics started when it was realized that strength calculations based on assuming perfect crystals were far too high because they ignored pre-existing flaws. Griffith[*] reasoned that a pre-existing crack could propagate under stress only if the release of elastic energy exceeded the work required to form the new fracture surfaces. However, his theory based on energy release predicted fracture strengths that were much lower than those measured experimentally. Orowan[†] realized that plastic work should be included in the term for the energy required to form a new fracture surface. With this correction, experiment and theory were finally brought into agreement. Irwin[‡] offered a new and entirely equivalent approach by concentrating on the stress states around the tip of a crack.

Theoretical Fracture Strength

Early estimates of the theoretical fracture strength of a crystal were made by considering the stress required to separate two planes of atoms. Figure 10.1 shows schematically how the stress might vary with separation. The attractive

---

[*] A. A. Griffith, *Phil. Trans. Roy. Soc. London, Ser A*, v. 221 (1920).
[†] E. Orowan, Fracture and Strength of Crystals, *Rep. Prog. Phys.*, v. 12 (1949).
[‡] G. R. Irwin, in *Fracturing of Metals*, ASM (1949).

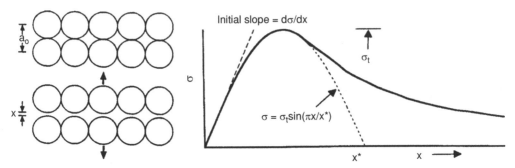

Figure 10.1. Schematic of the variation of normal stress with atomic separation and a sine wave approximation to the first part of the curve.

stress between two planes increases as they are separated, reaching a maximum that is the theoretical strength, σ_t, and then decaying to zero. The first part of the curve can be approximated by a sine wave of half wavelength, x^*,

$$\sigma = \sigma_t \sin(\pi x/x^*), \tag{10.1}$$

where x is the separation of the planes. Differentiating, $d\sigma/dx = \sigma_t(\pi/x^*) \cos(\pi x/x^*)$. At low values of x, $\cos(\pi x/x^*) \to 1$, so $d\sigma/dx \to \sigma_t(\pi/x^*)$. The engineering strain is $e = x/a_o$. Young's modulus, E, is the slope of the stress-strain curve, $d\sigma/d\varepsilon$, as $\varepsilon \to 0$:

$$E = d\sigma/d\varepsilon = (a_o/x^*)\pi\sigma_t. \tag{10.2}$$

Solving for the theoretical strength,

$$\sigma_t = Ex^*/(\pi a_o). \tag{10.3}$$

If it is assumed that $x^* = a_o$,

$$\sigma_t \approx E/\pi \tag{10.4}$$

Equation 10.4 predicts theoretical strengths that are very much higher than those that are observed (65 GPa vs. about 3 GPa for steel and 20 GPa vs. about 0.7 GPa for aluminum).

The assumption that $x^* = a_o$ can be avoided by equating the work per area to create two fracture surfaces to twice the specific surface energy (surface tension), γ. The work is the area under the curve in Figure 10.1,

$$2\gamma = \int_0^{x^*} \sigma\,dx = \int_0^{x^*} \sigma_t \sin(\pi x/x^*)dx = -(x^*/\pi)\sigma_t[\cos\pi - \cos 0] = (2x^*/\pi)\sigma_t, \tag{10.5}$$

so

$$x^* = \pi\gamma/\sigma_t. \tag{10.6}$$

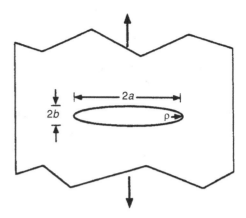

Figure 10.2. Internal crack in a plate approximated by an ellipse with major and minor radii of a and b.

Substituting Equation 10.6 into Equation 10.3,

$$\sigma_t = [(\pi\gamma/\sigma_t)/a_o]E/\pi, \tag{10.7}$$

$$\sigma_t = \sqrt{(\gamma E/a_o)}. \tag{10.8}$$

The predictions of Equations 10.8 and 10.4 are similar and much too large.

Stress Concentration

The reason that the theoretical predictions are high is that they ignore flaws, and all materials contain flaws. In the presence of a flaw, an externally applied stress is not uniformly distributed within the material. Discontinuities such as internal cracks and notches are stress concentrators. For example, the stress at the tip of the crack, σ_{max}, in a plate containing an elliptical crack (Figure 10.2) is given by

$$\sigma_{max} = \sigma_a(1 + 2a/b), \tag{10.9}$$

where σ_a is the externally applied stress. The term $(1 + 2a/b)$ is called the *stress concentration* factor. The radius of curvature, ρ, at the end of an ellipse is given by

$$\rho = b^2/a, \tag{10.10}$$

so $a/b = \sqrt{(a/\rho)}$. Substitution of Equation 10.10 into Equation 10.9 results in

$$\sigma_{max} = \sigma_a(1 + 2\sqrt{(a/\rho)}). \tag{10.11}$$

Because a/ρ is usually very large ($a/\rho \gg 1$), Equation 10.11 can be approximated by

$$\sigma_{max} = 2\sigma_a\sqrt{(a/\rho)}. \tag{10.12}$$

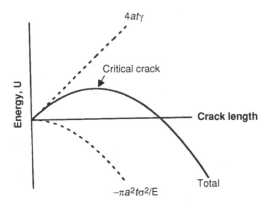

Figure 10.3. The effect of a crack's length on its energy. The elastic energy decreases and the surface energy increases. There is a critical crack length at which growth of the crack lowers the total energy.

It is possible to rationalize the difference between theoretical and measured fracture stresses in terms of Equation 10.12. Fracture will occur when the level of σ_{max} reaches the theoretical fracture strength, so the average stress, σ_a, at fracture is much lower than the theoretical value.

> **EXAMPLE PROBLEM #10.1:** Calculate the stress concentration at the tip of an elliptical crack having major and minor radii of $a = 100$ nm and $b = 1$ nm.
>
> *Solution:* $\rho = b^2/a = 10^{-18}/10^{-7} = 10^{-11}$. $\sigma_{max}/\sigma_a = 1 + 2\sqrt{(a/\rho)} = 1 + 2\sqrt{(10^{-7}/10^{-11})} = 201$.

Griffith and Orowan Theories

Griffith approached the subject of fracture by assuming that materials always have pre-existing cracks. He considered a large plate with a central crack under a remote stress, σ, and calculated the change of energy, ΔU, with crack size (Figure 10.3). There are two terms: One is the surface energy associated with the crack,

$$\Delta U_{surf} = 4at\gamma, \qquad (10.13)$$

where t is the plate thickness and $2a$ is the length of an internal crack. The other term is the decrease of stored elastic energy due to the presence of the crack,

$$\Delta U_{elast} = -\pi a^2 t\sigma^2/E. \qquad (10.14)$$

This term will not be derived here. However, to check its magnitude one can simplify the problem by assuming all the strain energy is lost inside a circle of diameter $2a$, and none outside. The energy/volume is $(1/2)\sigma\varepsilon = (1/2)\sigma^2/E$ and the volume is $\pi a^2 t$, so the total elastic strain energy would be $\Delta U_{elast} = -(1/2)\pi a^2 t\sigma^2/E$. The correct solution is just twice this. Note that the derivation of this assumes that plate width, w, is very large compared to a ($w \gg a$), and that the plate thickness, t, is small compared to a ($t \ll a$). For thick plates ($t \gg a$), plane-strain prevails and Equation 10.14 becomes $\Delta U_{elast} = -\pi a^2 t \sigma^2 (1 - \upsilon^2)/E$.

Combining Equations 10.13 and 10.14,

$$\Delta U_{total} = 4at\gamma - \pi a^2 t (\sigma^2/E). \tag{10.15}$$

This equation predicts that the energy of the system first increases with crack length and then decreases, as shown in Figure 10.3. Under a fixed stress, there is a critical crack size above which crack growth lowers the energy. This critical crack size can be found by differentiating Equation 10.15 with respect to a and setting it to zero, $d\Delta U_{total}/da = 4t\gamma - 2\pi at(\sigma^2/E) = 0$, so

$$\sigma = \sqrt{(2E\gamma/\pi a)}. \tag{10.16}$$

This is known as the Griffith criterion. A pre-existing crack of length greater than $2a$ will grow spontaneously when Equation 10.16 is satisfied. Griffith found reasonable agreement between this theory and experimental results on glass. However, the predicted stresses were very much too low for metals.

In this development, the plate is assumed to be thin enough relative to the crack length that there is no stress relaxation in the thickness direction, for example, plane-stress conditions prevail. On the other hand, if the plate is thick enough that there is complete strain relaxation in the thickness direction (plane-strain conditions), Equation 10.16 should be modified to

$$\sigma = \sqrt{2E\gamma/[(1 - \upsilon^2)\pi a]}. \tag{10.17}$$

Orowan proposed that the reason that the predictions of Equations 10.16 and 10.17 were too low for metals is because the energy expended in producing a new surface by fracture is not just the true surface energy. There is a thin layer of plastically deformed material at the fracture surface and the energy to cause this plastic deformation is much greater than the surface energy, γ. To account for this, Equation 10.16 is modified to

$$\sigma = \sqrt{(EG_c/\pi a)}, \tag{10.18}$$

where G_c replaces 2γ and includes the plastic work in generating the fracture surface.

Fracture Modes

There are three different modes of fracture, each having a different value of G_c. These modes are designated I, II, and III, as illustrated in Figure 10.4.

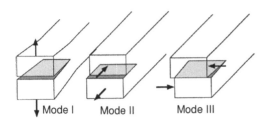

Figure 10.4. The three modes of cracking.

Mode I   Mode II   Mode III

In mode I fracture, the fracture plane is perpendicular to the normal force. This is what is occurs in tension tests of brittle materials. Mode II fractures occur under the action of a shear stress with the fracture propagating in the direction of shear. An example is the punching of a hole. Mode III fractures are also shear separations but here the fracture propagates perpendicular to the direction of shear. An example is cutting of paper with scissors.

Irwin's Fracture Analysis

Irwin noted that in a body under tension, the stress state around an infinitely sharp crack in a semi-infinite elastic solid is entirely described by*

$$\sigma_x = K_I/(2\pi r)^{1/2} \cos(\theta/2)[1 - \sin(\theta/2)\sin(3\theta/2)] \tag{10.19a}$$

$$\sigma_y = K_I/(2\pi r)^{1/2} \cos(\theta/2)[1 + \sin(\theta/2)\sin(3\theta/2)] \tag{10.19b}$$

$$\tau_{xy} = K_I/(2\pi r)^{1/2} \cos(\theta/2)\sin(\theta/2)\cos(3\theta/2) \tag{10.19c}$$

$$\sigma_z = \upsilon(\sigma_x + \sigma_y) \text{ for plane strain } (\varepsilon_z = 0) \text{ and } \sigma_z = 0 \text{ for plane stress} \tag{10.19d}$$

$$\tau_{yz} = \tau_{zx} = 0, \tag{10.19e}$$

where θ and r are the coordinates relative to the crack tip (Figure 10.5) and K_I is the stress intensity factor,[†] which in a semi-infinite body is given by

$$K_I = \sigma(\pi a)^{1/2}. \tag{10.20}$$

Here σ is the applied stress. For finite specimens, $K_I = f\sigma(\pi a)^{1/2}$, where f depends on the specimen geometry and is usually a little greater than 1 for small cracks. Figure 10.5 shows how f varies with the ratio of crack length, a, to the specimen width, w.

The Westergaard equations predict that the local stresses, σ_x and σ_y, are infinite at the crack tip and decrease with distance from the crack, as shown in Figure 10.6. Of course, the infinite stress prediction is unrealistic. The material will yield wherever σ_y is predicted to be greater than the material's yield strength, so yielding limits the actual stress near the crack tip ($\sigma_y \leq Y$).

* These equations are attributed to Westergaard, *Trans. ASME, J. Appl. Mech.* v. 61 (1939).
† The stress intensity factor should not be confused with the stress concentration factor.

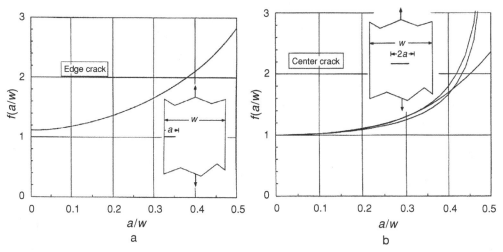

Figure 10.5. Variation of f with a/w for edge and center cracks. The three lines for the center crack are from three different mathematical approximations.

In Irwin's analysis, fracture occurs when K_I reaches a critical value, K_{Ic}, which is a material property. This predicts that the fracture stress, σ_f, is

$$\sigma_f = K_{Ic}/[f(\pi a)^{1/2}]. \tag{10.21}$$

Comparison with Equation 10.18 shows that for $f = 1$,

$$K_{Ic} = \sqrt{(EG_C)}. \tag{10.22}$$

EXAMPLE PROBLEM #10.2: A 6 ft long steel strut, 2.0 in. wide and 0.25 in. thick, is designed to carry an 80,000 lb tension load. Assume that the steel has a toughness of 40 ksi√in. What is the longest edge crack that will not cause failure?

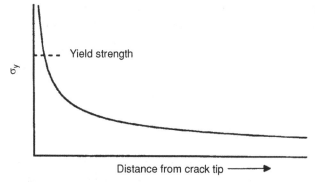

Figure 10.6. Stress distribution ahead of a crack. The Westergaard equations predict an infinite stress at the crack tip, but the stress there can be no greater than the yield strength.

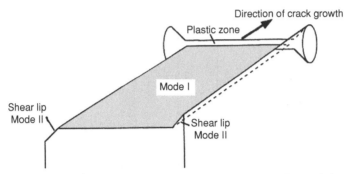

Figure 10.7. A three-dimensional sketch showing the shape of the plastic zone and the shear lip formed at the edge of the plate.

Solution: According to Equation 10.21, failure will occur if $a = (K_{Ic}/f\sigma)^2/\pi$. To solve this problem, we must first guess a value for f. As a first guess let $f = 1.2$. Substituting $f = 1.2$, $K_{Ic} = 40$ ksi$\sqrt{\text{in}}$. and $\sigma = 80{,}000$ lb/$(2.0 \times 0.25$ in.$^2) = 160$ ksi, $a = 0.014$ in. Now checking to see whether the guess of $f = 1.2$ was reasonable, $a/w = 0.08/2 = 0.04$. From Figure 10.5a, $f = 1.12$. Repeating the calculation with $f = 1.12$, $a = 0.016$. The value $f = 1.12$ is reasonable for $a = 0.016$.

Plastic Zone Size

The plastic work involved in the yielding of material near the crack tip as the crack advances is responsible for the energy absorption. Figure 10.7 is a three-dimensional sketch of the plastic zone shape. Plane-strain ($\varepsilon_z = 0$) is characteristic of the interior where adjacent material prevents lateral contraction. At the surface where plane stress ($\sigma_z = 0$) prevails, the fracture corresponds to mode II rather than mode I. The energy absorption per area is greater here because of the larger volume of deforming material in the plane stress region.

The sizes and shapes of the plastic zones calculated for plane strain ($\varepsilon_z = 0$) and plane stress ($\sigma_z = 0$) conditions are shown in Figure 10.8. It is customary to characterize the size of the plastic zone by a radius, r_p. For plane stress,

$$r_p = (K_{Ic}/Y)^2/(2\pi). \tag{10.23}$$

For plane strain,

$$r_p = (K_{Ic}/Y)^2/(6\pi). \tag{10.25}$$

There is no stress, σ_z, normal to the surface of the specimen, so plane stress prevails at the surface with a characteristic toughness K_{IIc} that is greater than K_{Ic}. There is a transition from plane stress to plane strain with increasing distance from the surface. To obtain valid K_{Ic} data, the thickness of the specimen

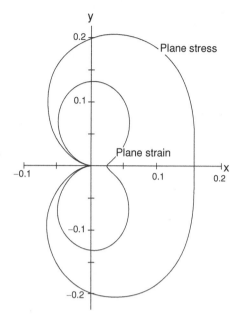

Figure 10.8. Plots of the plastic zones associated with plane stress and plane strain. Dimensions are in units of $(K_I/Y)^2$. The curve for plane strain was calculated for $\upsilon = 0.3$. Note that the "radii" of the plane-stress and plane-strain zones are conventionally taken as $(K_I/Y)^2/(2\pi) = 0.159$ and $(K_I/Y)^2/(6\pi) = 0.053$, respectively.

should be much greater than the radius of the plastic zone for plane stress. Specifications require that

$$t \geq 2.5(K_{Ic}/Y)^2. \tag{10.24}$$

This assures that the fraction of the fracture surface failing under plane-stress conditions, $2r_p/t$, equals or is less than $[2(K_{Ic}/Y)^2/(2\pi)]/[2.5(K_{Ic}/Y)^2] = 12.7\%$.

EXAMPLE PROBLEM #10.3: Using stress field in Equations 10.19 for plane-stress ($\sigma_z = 0$), calculate the values of r at $\theta = 0°$ and $90°$ at which the von Mises yield criterion is satisfied. Express the values of r in terms of (K_{Ic}/Y).

Solution: According to Equations 10.19, for $\theta = 0$, $\sigma_x = \sigma_y = K_{Ic}/\sqrt{(2\pi r)}$, $\sigma_z = 0$, $\tau_{yz} = \tau_{zx} = \tau_{xy} = 0$. Substituting into the von Mises criterion, $\sigma_x = \sigma_y = Y$ so $Y = K_{Ic}/\sqrt{(2\pi r)}$. Solving for r, $r = (K_{Ic}/Y)^2/(2\pi) = 0.159(K_{Ic}/Y)^2$.

For $\theta = 90°$, $\theta/2 = 45°$, $\cos(\theta/2) = \sin(\theta/2) = \sin(3\theta/2) = -\cos(3\theta/2) = 1/\sqrt{2}$. According to Equations 10.19, $\sigma_x = K_{Ic}/\sqrt{(2\pi r)}[1/(2\sqrt{2})]$, $\sigma_y = K_{Ic}/\sqrt{(2\pi r)}[3/(2\sqrt{2})] = 3\sigma_x$, $\sigma_z = 0$, $\tau_{xy} = K_{Ic}/\sqrt{(2\pi r)}[-1/(2\sqrt{2})] = -\sigma_x$, $\tau_{yz} = \tau_{zx} = 0$. Substituting into the von Mises criterion, $2Y^2 = (3\sigma_x - 0)^2 + (0 - \sigma_x)^2 + (\sigma_x - 3\sigma_x)^2 + 6(\sigma_x)^2 = 20\sigma_x^2$ so $\sigma_x^2 = 2Y^2/20 = Y^2/10$. Now substituting $\sigma_x^2 = K_{Ic}^2/(2\pi r)[1/(2\sqrt{2})]^2 = Y^2/10$,

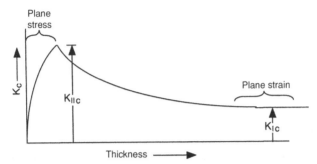

Figure 10.9. The dependence of K_c on thickness. For very thick plates, the effect of mode II in the surface region is negligible, so K_c approaches K_{Ic}. For thinner sheets, more of the fracture surface is characterized by plane-stress conditions, so K_c approaches K_{IIc}. For even thinner sheets, K_c decreases with thickness because the plastic volume decreases.

$r = [10/(16\pi)](K_{Ic}/Y)^2 = 0.199(K_{Ic}/Y)^2$. Compare these values with Figure 10.9 fo $\theta = 0°$ and $\theta = 90°$ for plane stress conditions.

The overall fracture toughness (critical stress intensity), K_c, depends on the relative sizes of the plane-stress and plane-strain zones, and therefore depends on the specimen thickness, as sketched in Figure 10.9.

Thin Sheets

For thick sheets, K_c decreases with thickness because a lower fraction of the fracture surface is in mode II. However, if the sheet thickness is less than twice r_p for plane stress, the entire fracture surface fails in mode II. In this case, the failure is by through-thickness necking (Figure 10.10). The volume of the plastic region equals t^2L, where t is the specimen thickness and L is the length of the crack. Because the area of the fracture is tL, the plastic work per area is proportional to the thickness, t. Therefore, K_c is proportional to $\sqrt{t}$. Very thin sheets tear at surprisingly low stresses. Plates made of laminating sheets have lower toughness than monolithic ones (see Figure 10.11).

EXAMPLE PROBLEM #10.4: Derive expressions that show how K_c and G_c depend on thickness for thin sheets that fail by plane-stress fracture.

Solution: The fracture energy per area is $G_c = U/A = U/(tL)$, where U is the plastic work and L is the length of the fracture. Assume that necks in different thickness specimens are geometrically similar. Then the necked volume $= Ct^2L$ so $U = C'Lt^2$, where C and C' are constants. Substituting, $G_c = U/(tL) = C'Lt^2/(tL) = C't$, so G_c is inversely proportional to t. $K_c = \sqrt{(G_cE)} = \sqrt{(C'tE)}$, so K_c is inversely proportional to the square root of thickness.

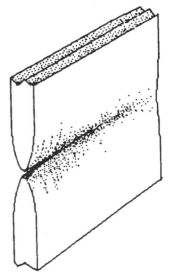

Figure 10.10. Sketch of a sheet failure by through-thickness necking.

Metallurgical Variables

Most metallurgical variations responsible for increasing strength, such as increased carbon content of steels, alloying elements in solid solution, cold working, and martensitic hardening, tend to decrease toughness. The sole exception seems to be decreased grain size, which increases both strength and toughness. Figure 10.12 shows the decrease of toughness with strength for 4140 steels that had been tempered to different yield strengths. There is a similar correlation between strength and toughness for aluminum alloys. Other factors such as the presence of inclusions also affect K_{Ic}.

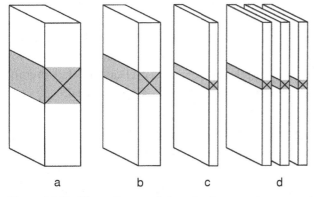

a   b   c   d

Figure 10.11. The volume of the plastic zone is proportional to t^2, so the energy absorbed per area of fracture surface decreases as t decreases. The toughness of a laminate (D) is less than that of a monolithic sheet of the same total thickness (A) because the plastic volume is smaller.

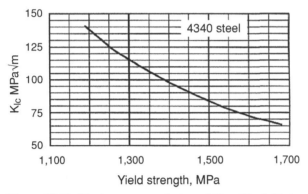

Figure 10.12. The inverse correlation of K_{Ic} with yield strength for 4340 steel.

Fracture Mechanics in Design

Engineering design depends on the size of the largest possible crack that might exist in a stressed part or component. Various techniques are used to inspect for pre-existing cracks. These include ultrasonics, x-rays, magnetic inspection, and dye penetrants. For each technique, there is a limit to how small a crack can be reliably detected. Safe design is based on assuming the presence of cracks of the largest size that cannot be detected with 100% certainty. For example, if that inspection technique cannot assure detection of edge cracks smaller than 1 mm, the designer must assume $a = 1$ mm. Then the permissible stress is calculated using Equation 10.21 and applying an appropriate safety factor. The design should also assure that the component does not yield plastically, so the permissible stress should not exceed the yield strength multiplied by the safety factor. If the stress calculated from Equation 10.21 exceeds the yield strength, failure from accidental overload will occur by plastic yielding rather than fracture. However, yielding is usually regarded as less dangerous than fracture.

Even if there are no pre-existing cracks larger than those assumed in the design, "safe failure" is still not assured because smaller cracks may grow by fatigue during service. In critical applications, such as cyclically loaded aircraft components, periodic inspection during the life of a part is required.

EXAMPLE PROBLEM #10.5: A support is to be made from 4340 steel. The steel may be tempered to different yield strengths. The correlation between strength and toughness was shown in Figure 10.13. Assume that $f = 1.1$ for the support geometry and that nondestructive inspection can detect all edge cracks of size $a = 2$ mm or larger.

A. If a 4340 steel of $K_{Ic} = 120$ MPa$\sqrt{m}$ is used, what is the largest stress that will not cause either yielding or fracture? If the support is overloaded, will it fail by yielding or by fracture?

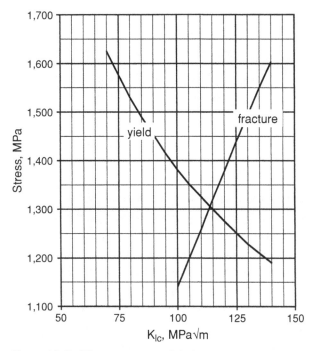

Figure 10.13. The greatest possible load is reached when $\sigma_f = Y = 1305$ MPa.

B. By selecting a 4340 steel of a different yield strength, find the yield strength that will allow the largest stress without either yielding or fracture.

Solution: A. From Figure 10.13, for $K_{Ic} = 120$ MPa$\sqrt{m}$. $Y = 1275$ MPa. Using Equation 10.21, $\sigma_f = K_{Ic}/[f\sqrt{(\pi a)}] = 120$ MPa$/[1.1\sqrt{(0.002m.\pi)}] = 1376$ MPa. This is greater than the yield strength, so it will fail by yielding when $\sigma = 1275$ MPa.

B. For the greatest load carrying capacity, yielding and fracture should occur at the same stress level. Figure 10.13 is a plot of calculated values of σ_f for several different levels of K_{Ic}. The yield strength vs. K_{Ic} data from Figure 10.13 is replotted on the same axes. The intersection of two curves indicates that the optimum value of Y is about 1305 MPa.

Compact Tensile Specimens

The *compact tensile specimen* (Figure 10.14) is often used for measuring K_{Ic}. The specimen is loaded by pins inserted into the holes. The dimensions are selected so that loading of the specimen will be entirely elastic except for the plastic zone at the crack tip. With the requirement that $a \geq 2.5(K_{Ic}/Y)^2$, the radius of the plane-stress plastic zone at the surface $r_p = (K_{Ic}/Y)^2/(2\pi)$ is $\leq$

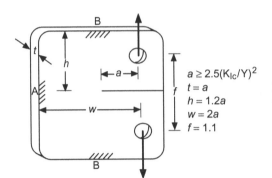

Figure 10.14. Compact tensile specimen used to measure K_{Ic}.

$a \geq 2.5(K_{Ic}/Y)^2$
$t = a$
$h = 1.2a$
$w = 2a$
$f = 1.1$

$a/5\pi$. Mode II extends inward from each surface a distance of r_p, so mode II will prevail over less than $a/(2.5\pi)$. If the thickness is equal to a, this amounts to $1/(2.5\pi) = 12.7\%$ of the fracture surface. The requirement that $w/a = 2$ insures that a plastic hinge will not develop at the back of the specimen (region A). Similarly, if h were less than $1.2a$, regions B would yield in bending.

The initial crack is made by machining. Then cyclic loading is applied to cause the crack grow and sharpen by fatigue. Finally, the fracture toughness is measured by loading the specimen until the crack grows. The load at this point is noted and K_{Ic} is calculated from Equation 10.19, taking σ as the load divided by w and t.

EXAMPLE PROBLEM #10.6: A medium-carbon steel has a yield strength of 250 MPa and a fracture toughness of 80 MPa$\sqrt{\text{m}}$. How thick would a compact tensile specimen have to be for valid plane-strain fracture testing?

Solution: The specimen thickness should be at least $2.5(K_{Ic}/Y)^2 = 2.5(80/250)^2 = 0.26$ m, or about 10 inches. (This would be impossible unless a very thick plate and a very large testing machine were available.)

The *J*-Integral

Linear elastic behavior was been assumed in the development of Equations 10.19 that form the basis for the treatment up to this point. For relatively tough materials, very large specimens are required to assure elastic behavior. Often materials are not available in the thicknesses needed. The *J*-integral offers a method of evaluating the toughness of a material that undergoes a nonlinear behavior during loading. Strictly, it should be applied only for nonlinear elastic behavior, but it has been shown that the errors caused by a limited amount of plastic deformation are not large.

J is the work done on a material per area of fracture. It is the area between the loading path and the unloading path after the fracture area divided by the increase of fracture area, Δa. If there has been no plastic deformation, the unloading will return to the origin as shown in Figure 10.15A. This area can

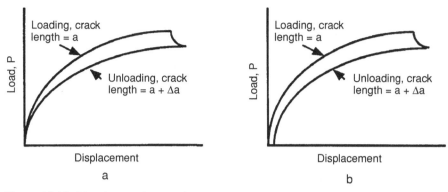

Figure 10.15. J is taken as the area between the loading and unloading curves. (A) For fully elastic behavior, the unloading curve returns to the origin. (B) If some plasticity accompanies the crack propagation, the unloading curve will not return to the origin.

be measured by unloading or by testing a specimen with a longer initial crack. The value of J is equal to G_c,

$$J_{Ic} = G_{Ic} = K_{Ic}^2/[E/(1 - \upsilon^2)].$$ (10.31)

However, if there has been plastic deformation accompanying the crack propagation, the unloading curve will not return to the origin, as shown in Figure 10.15b. In this case, J is still taken as the area between the loading and unloading curves, and J should be somewhat larger than G_c. Figure 10.16 is a comparison of experimentally determined values of J_{Ic} and K_{Ic}.

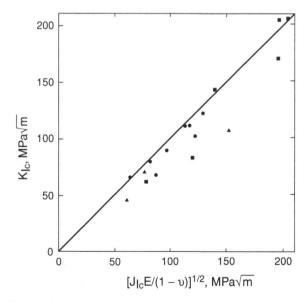

Figure 10.16. Comparison of experimentally determined values of J_{Ic} and K_{Ic} for several steels. Adapted from R. W. Hertzberg, *Deformation and Fracture and Fracture Mechanics of Engineering Materials*, 4th Ed., John Wiley (1995).

There are other approaches to measuring the fracture toughness of tough materials, including measurements based on the crack opening displacement (COD). These methods will not be treated here.

Notes

Alan A. Griffith is considered to be the father of fracture mechanics. He received his B. Eng., M. Eng. and D. Eng. degrees from the University of Liverpool. In 1915, he was employed by the Royal Aircraft Establishment (then the Royal Aircraft Factory), where he had a brilliant career. He did pioneering work on turbines that led to the early development of the jet engine. In his work on aircraft structures, he became interested in the stress concentrations caused by notches and scratches. He realized that if the curvature at the end of a crack were of molecular dimensions, the stress concentration would reduce strengths far below those that were actually observed. Griffith measured the fracture strength of glass fibers that naturally contained small cracks. He found reasonable agreement with his theory. Such agreement would not have been found had he worked with metals or other materials in which there is energy absorbed by the plastic zone associated with crack propagation. His classic paper, in *Phil. Trans. Roy. Soc London*, Ser. A, v. 221A (1923), has been reprinted with commentary in *Trans. ASM* v. 61 (1968).

REFERENCES

R. W. Hertzberg, *Deformation and Fracture Mechanics of Engineering Materials*, 4th ed., John Wiley (1996).
T. H. Courtney, *Mechanical Behavior of Materials*, McGraw-Hill (1990).
G. E. Dieter, *Mechanical Metallurgy*, 3rd ed. McGraw-Hill (1986).
N. E. Dowling, *Mechanical Behavior of Materials*, Prentice Hall (1993).

Problems

1. Class 20 and class 60 gray cast irons have tensile strengths of about 20 ksi and 60 ksi, respectively. Assuming that the fractures start from graphite flakes and that the flakes act as pre-existing cracks, use the concepts of the Griffith analysis to predict the ratio of the average graphite flake sizes of the two cast irons.

2. A wing panel of a supersonic aircraft is made from a titanium alloy that has a yield strength 1035 MPa and toughness of $K_{Ic} = 55 \text{MPa}\sqrt{m}$. It is 3.0 mm thick, 2.40 m long, and 2.40 m wide. In service, it is subjected to a cyclic stress of ± 700 MPa, which is not enough to cause yielding but does cause gradual crack growth of a pre-existing crack normal to the loading direction at the edge of the panel. Assume that the crack is initially 0.5 mm

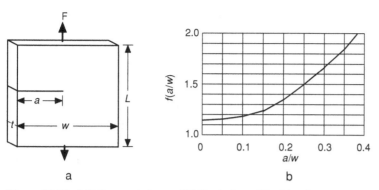

Figure 10.17. (A) Support shape. (B) Variation of f with a/w.

long and grows at a constant rate of $da/dN = 120$. Calculate the number of cycles to failure.

3. The support in Figure 10.17a is to be constructed from a 4340 steel plate tempered at 800F. The yield strength of the steel is 228 ksi and its value of K_{Ic} is 51 ksi$\sqrt{(\text{in})}$. The width of the support, w, is 4 in., the length, L, is 36 in. and the thickness, t, is 0.25 in. Figure 10.17b gives $f(a/w)$ in the equation $\sigma = K_{Ic}/[f(a/w)\sqrt{(\pi a)}]$.

   A. If the crack length, a, is small enough, will the support yield before it fractures? What is the size of the largest crack for which this is true (i.e., what is the largest value of a for which general yielding will precede fracture)? Assume that any fracture would be in mode I (plane strain). Discuss critically the assumption of mode I.

   B. If it is guaranteed that there are no cracks longer than a length equal to 80% of that in A, does this assure that failure will occur by yielding rather than fracture? Explain briefly.

4. For 4340 steel, the fracture toughness and yield strength depend on the prior heat treatment, as shown in Figure 10.13. As yield strength is increased, the fracture toughness decreases. A pipeline is to be built of this steel and to minimize the wall thickness of the pipe, the stress in the walls should be as large as possible without either fracture or yielding. Inspection techniques ensure that there are no cracks longer than 2 mm ($a = 1$ mm). What level of yield strength should be specified? Assume a geometric constant of $f = 1.15$.

5. A steel plate, 10 ft long, 0.25 in. thick, and 6 in. wide, is loaded under a stress of 50 ksi. The steel has a yield strength of 95 ksi and a fracture toughness of 112 ksi$\sqrt{(\text{in})}$. There is a central crack perpendicular to the 10-ft dimension.

   A. How long would the crack have to be for failure of the plate under the stress?

B. If there is an accidental overload (i.e., the stress rises above the specified 50 ksi), the plate might fail by either yielding or by fracture depending on the crack size. If the designer wants to be sure that an accidental overload would result in yielding rather than fracture, what limitations must be placed on the crack size?

6. A structural member is made from a steel that has a $K_{Ic} = 180\text{MP}\sqrt{m}$ and a yield strength of 1050 MPa. In service, it should neither break nor deform plastically as either would be considered a failure. Assume $f = 1.0$. If there is a pre-existing surface crack of a $= 4$ mm, at what stress will the structural member fail? Will they fail by yielding or fracture?

7. An estimate of the effective strain, ε, in a plane-strain fracture surface can be made in the following way. Assume that the material is not work-hardened and the effective strain is constant throughout the plastic zone, so the plastic work per volume is $Y\varepsilon$. Assume the depth of the plastic zone is given by the equation $r_p = (K_I/Y)^2/(6\pi)$ and the strained volume is $2r_p A$. Derive an expression for the plastic strain, ε, associated with running of a plane-strain fracture. (Realize that G_c is the plastic work per crack area, and that K_{Ic} and G_c are related by Equation 10.22.)

11 Fatigue

Introduction

It has been estimated that 90% of all service failures of metal parts are caused by *fatigue*. A fatigue failure is one that occurs under a cyclic or alternating stress of an amplitude that would not cause failure if applied only once. Aircraft are particularly sensitive to fatigue. Automobile parts such as axles, transmission parts, and suspension systems may fail by fatigue. Turbine blades, bridges, and ships are other examples. Fatigue requires cyclic loading, tensile stresses, and plastic strain on each cycle. If any of these are missing, there will be no failure. The fact that a material fails after a number of cycles indicates that some permanent change must occur on every cycle. Each cycle must produce some plastic deformation, even though it may be very small. Metals and polymers fail by fatigue. Fatigue failures of ceramics are rare because there seldom is plastic deformation.

There are three stages of fatigue. The first is nucleation of a crack by small amounts of inhomogeneous plastic deformation at a microscopic level. The second is the slow growth of these cracks by cyclic stressing. Finally, sudden fracture occurs when the cracks reach a critical size.

Surface Observations

Often visual examination of a fatigue fracture surface will reveal *clamshell* or *beach markings*, as shown in Figure 11.1. These marks indicate the position of the crack front at some stage during the fatigue life. The initiation site of the crack can easily be located by examining these marks. The distance between these markings does not represent the distance that the crack propagated in one cycle. Rather, each mark corresponds to some change during the cyclic loading history, a period of time that allowed corrosion or a change in the stress amplitude. When the crack has progressed far enough, the remaining portion of the cross-section may fail in one last cycle by either brittle or ductile

Figure 11.1. Typical clamshell markings on a fatigue fracture surface of a shaft. The fracture started at the left side of the bar and progressed to the right, where final failure occurred in a single cycle. Courtesy of W.H. Durrant.

fracture. The final fracture surface may represent the major part or only a very small portion of the total fracture surface depending on the material, its toughness, and the loading conditions.

Microscopic examination of a fracture surface often reveals markings on a much finer scale. These are called *striations* and they do represent the position of the crack front at each cycle (Figure 11.2). The distance between striations is the distance advanced by the crack during one cycle. Sometimes striations cannot be observed because they are damaged when the crack closes.

A careful microscopic examination of the exterior surface of the specimen after cyclic stressing will usually reveal a roughening even before any cracks have formed. Under high magnification, intrusions and extrusions are often apparent, as shown in Figure 11.3. These intrusions and extrusions are the result of the slip of one set of planes during the compression half-cycle and

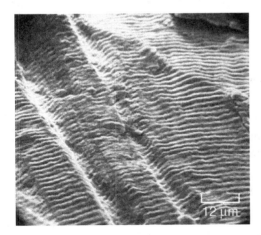

Figure 11.2. SEM picture of fatigue striations on a fracture surface of type 304 stainless steel. From *Metals Handbook, v. 9*, 8th ed., ASM (1974).

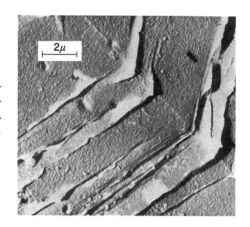

Figure 11.3. Intrusions and extrusions at the surface formed by cyclic deformation. These correspond to persistent slip bands beneath the surface. From A. Cottrell and D. Hull, *Proc. Roy. Soc. (London)* v. A242 (1957).

the slip on a different set of planes during the tensile half-cycle (Figure 11.4). *Persistent slip bands* beneath the surface are associated with these intrusions and extrusions. Fatigue cracks initiate at the intrusions and grow inward along the persistent slip bands.

Nomenclature

Most fatigue experiments involve alternate tensile and compressive stresses, often applied by cyclic bending. In this case, the mean stress is zero. However, in service materials may be subjected to cyclic stresses that are superimposed on a steady-state stress. Figure 11.5 shows this together with various terms

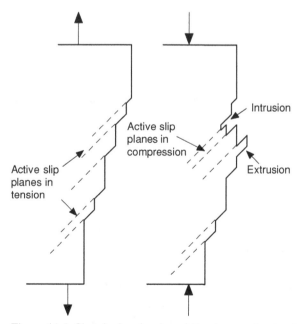

Figure 11.4. Sketch showing how intrusions and extrusions can develop if slip occurs on different planes during the tension and compression portions of the loading.

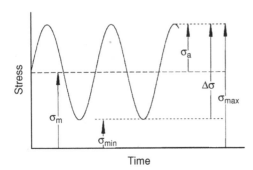

Figure 11.5. Schematic of cyclic stresses, illustrating several terms.

used to define the stresses. The mean stress, σ_m, is defined as

$$\sigma_m = (\sigma_{max} + \sigma_{min})/2. \tag{11.1}$$

The amplitude, σ_a, is

$$\sigma_a = (\sigma_{max} - \sigma_{min})/2, \tag{11.2}$$

and the range, $\Delta\sigma$, is

$$\Delta\sigma = (\sigma_{max} - \sigma_{min}) = 2\sigma_a. \tag{11.3}$$

Also, the ratio of maximum and minimum stresses is

$$R = \sigma_{min}/\sigma_{max}, \tag{11.4}$$

which can be expressed as

$$R = (\sigma_m - \Delta\sigma/2)/(\sigma_m + \Delta\sigma/2). \tag{11.5}$$

For completely reversed loading, $R = -1$, and for tension-release, $R = 0$. Static tensile loading corresponds to $R = 1$.

Although Figure 11.5 is drawn with sinusoidal stress waves, the actual wave shape is of little or no importance. Frequency is also unimportant unless it is so high that heat cannot be dissipated and the specimen heats up or so low that creep occurs during each cycle.

S-N Curves

Most fatigue data are presented in the form as *S-N* curves, which are plots of the cyclic stress amplitude ($S = \sigma_a$) vs. the number of cycles to failure (N), with N conventionally plotted on a logarithmic scale. Usually, *S-N* curves are for tests in which the mean stress (σ_m) is zero. Figures 11.6 and 11.7 show the *S-N* curves for a 4340 steel and a 7075 aluminum alloy. The number of cycles to cause failure decreases as the stress amplitude increases.

For low-carbon steels and other materials that strain age, there is a stress amplitude (*endurance limit* or *fatigue limit*) below which failure will never

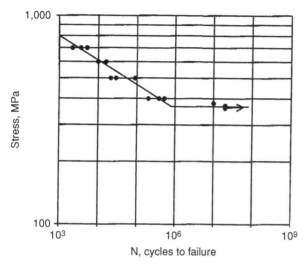

Figure 11.6. The *S-N* curve for annealed 4340 steel. Typically, the break in the curve for a material with an fatigue limit occurs at about 10^6 cycles. The points with arrows are for tests stopped before failure.

occur. The break in the *S-N* curve occurs at about 10^6 cycles, as shown in Figure 11.6. However, many materials, such as aluminum alloys (Figure 11.7), have no true fatigue limit. The stress amplitude for failure continues to decrease, even at a very large number of cycles. In this case, the *fatigue strength* is often defined as the stress amplitude at which failure will occur in 10^7 cycles.

If the *S-N* curve is plotted as $\log(S)$ against $\log(N)$, as in Figures 11.6 and 11.7, a straight line often results for $N < 10^6$. In this case, the relation may be

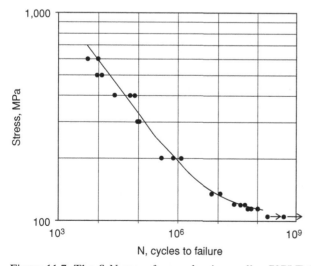

Figure 11.7. The *S-N* curve for an aluminum alloy 7075 T-6. Note that there is no true fatigue limit.

expressed as

$$S = AN^{-b},\tag{11.6}$$

where N is the number of cycles to failure. The constant, A, is approximately equal to the tensile strength.

EXAMPLE PROBLEM #11.1: Use the initial linear portion of the S-N curve for aluminum alloy 7075 T-6 in Figure 11.7 to find the values of A and b in Equation 11.6.

Solution: For two points on the linear section, $S_1/S_2 = (N_1/N_2)^{-b}$, so

$$-b = \ln(S_1/S_2)/\ln(N_1/(N_2)). \text{ Substituting } S_1 = 600 \text{ MPa at } N_1 = 10^4$$

and

$$S_2 = 200 \text{ MPa at } N_2 = 10^6, -b = \ln(3)/\ln(10^{-2}) = 0.24$$

$A = (S)/(N)^{-b}$. Substituting $S_2 = 200$ MPa at $N_2 = 10^6$ and $b = 0.24$, A = 5400 MPa.

Effect of Mean Stress

Most S-N curves are from experiments in which the mean stress was zero. However, under service conditions the mean stress usually is not zero. Several simple engineering approaches to predicting fatigue behavior when the stress cycles about a mean stress have been proposed. Goodman suggested that

$$\sigma_a = \sigma_e[1 - \sigma_m/UTS],\tag{11.7}$$

where σ_a is stress amplitude corresponding to a certain life, σ_m is the mean stress, and σ_e is the stress amplitude that would give the same life if σ_m were zero. UTS is the ultimate tensile strength.

Soderberg proposed a more conservative relation,

$$\sigma_a = \sigma_e[1 - \sigma_m/YS].\tag{11.8}$$

where YS is the yield strength.

Gerber proposed a less conservative relation,

$$\sigma_a = \sigma_e[1 - (\sigma_m/UTS)^2].\tag{11.9}$$

These are plotted in Figure 11.8. Any combination of σ_m and σ_a outside this region will result in fatigue failure.

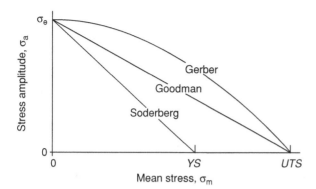

Figure 11.8. A plot representing the Goodman, Soderberg, and Gerber relations for the effect of mean stress on the stress amplitude for fatigue failure.

These relations may also be represented as plots of σ_{min} and σ_{max} vs. σ_m. Figure 11.9 is such a representation for the Goodman relation. In this plot, cycling between the σ_{min} and σ_{max} will not result in fatigue. Note that the permissible cyclic stress amplitude, σ_a, decreases as the mean stress, σ_m, increases, reaching zero at the tensile strength.

The conditions leading to yielding may be added to the Goodman diagram, as in Figure 11.10.

EXAMPLE PROBLEM #11.2: A bar of steel having a yield strength of 40 ksi, a tensile strength of 65 ksi, and an endurance limit of 30 ksi is subjected to a cyclic loading. Using a modified Goodman diagram, predict whether the material has an infinite fatigue life or whether it will fail by yielding

Figure 11.9. An alternative representation of the Goodman relation. Cycling outside of the lines σ_{min} and σ_{max} will result in failure.

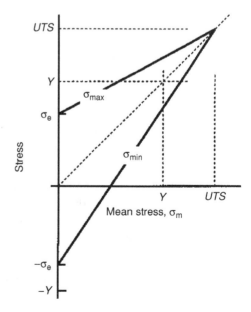

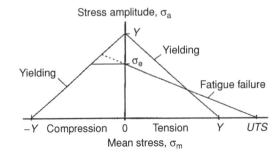

Figure 11.10. Modified Goodman diagram showing the effect of mean stress on failure by fatigue and yielding. Combinations of σ_a and σ_m above the lines $-YS$ to YS and YS to YS will result in yielding, whereas combinations of σ_a and σ_m above the line σ_e to UTS will result in eventual fatigue failure.

or fatigue for the following cases. The cyclic stress is (A) between 0 and 36 ksi, (B) between –27 ksi and +37, and (C) between 14 ksi ± 32 ksi.

Solution: Draw a Goodman diagram and plot each case on the diagram. For $A, \sigma_m = 18, \sigma_a = 18$, predict an infinite life; for $B, \sigma_m = 5, \sigma_a = 32$, predict fatigue failure without yielding; for $C, \sigma_m = 14, \sigma_a = 32$, predict yielding and fatigue failure (see Figure 11.11).

The Palmgren-Miner Rule

The *S-N* curve describes fatigue behavior at a constant stress amplitude, but often in service the cyclic amplitude varies during the life of a part. There may be periods of high stress amplitude followed by periods of low amplitude, or vice versa. This is certainly true of the springs of an automobile that sometimes drives on smooth roads and sometimes over potholes. Palmgren[*] and Miner[†] suggested a simple approximate rule for analyzing fatigue life under these conditions. The rule is that fatigue failure will occur when $\Sigma(n_i/N_i) = 1$, or

$$n_1/N_1 + n_2/N_2 + n_3/N_3 + \cdots = 1, \tag{11.10}$$

where n_i is the number of cycles applied at an amplitude, σ_{ai}, and N_i is the number of cycles that would cause failure at that amplitude. The term n_i/N_i represents the fraction of the life consumed by n_i cycles at σ_{ai}. When $\Sigma(n_i/N_i) = 1$, the entire life is consumed. This rule predicts that the fraction of the fatigue life consumed by n_i cycles at a given stress amplitude depends on the total life, N_i, at that stress amplitude.

According to this approximate rule, the order of cycling is of no importance. Yet experiments have shown that the life is shorter than predicted by Equation 11.10 if the amplitudes of the initial cycles are larger than the later ones. Likewise, if the initial cycles are of lower amplitude than the later ones, the life will exceed the predictions of the Palmgren-Miner rule. For steels,

[*] A. Palmgren, *Z.Verien Deutscher Ingenieur*, v. 68 (1924).
[†] M. A. Miner, *Trans. ASME., J. Appl. Mech.* v. 67 (1945).

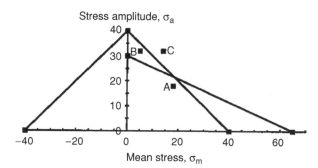

Figure 11.11. Modified Goodman diagram for example Problem 18.2.

cycling at stresses below the endurance limit will promote longer lives. This practice is called *coaxing*.

> **EXAMPLE PROBLEM #11.3:** A part made from the 7075-T4 aluminum alloy in Figure 11.7 has been subjected to 200,000 cycles of 250 MPa and 40,000 cycles of 300 MPa stress amplitude. According to Miner's rule, how many additional cycles of at 200 MPa can it withstand before failing?

> *Solution:* Using the results of Example #11.1, $N = (S/5400)^{-1/0.24}$ $N_{250} = 3.63 \times 10^5$, $N_{300} = 1.70 \times 10^5$, $N_{200} = 10^6$. The remaining life at 200 MPa is $n = N_{200}(1 - n_{250}/N_{250} - n_{300}/N_{300}) = 10^6[1 - (2 \times 10^5/3.63 \times 10^5) - (4 \times 10^4/1.70 \times 10^5)] = 0.21 \times 10^6$ cycles.

Stress Concentration

At an abrupt change in cross-section, the local stress can be much greater than the nominal stress. The *theoretical stress concentration factor*, K_t, is the ratio of the maximum local stress to the nominal stress, calculated by assuming elastic behavior. In Chapter 10, the stress concentration factors for elliptical holes were given by Equations 10.11 and 10.12 in semi-infinite plates. Figure 11.12 shows calculated values of K_t for circular holes and round notches in finite plates. Stress concentrators reduce fatigue strengths. Therefore, avoidance of such stress risers greatly lowers the likelihood of fatigue failure. However, the effect of notches on fatigue strength is not as great as would be expected by assuming that the actual stress was K_t times the nominal stress. Plastic deformation at the base of a notch reduces the actual stress there. How much the stress is reduced varies from material to material. The role of the material can be accounted for by a *notch sensitivity factor*, q, defined as

$$q = (K_f - 1)/(K_t - 1), \qquad (11.11)$$

where K_f is the *notch fatigue factor*, defined as the unnotched fatigue strength divided by the notched fatigue strength. If a notch causes no reduction

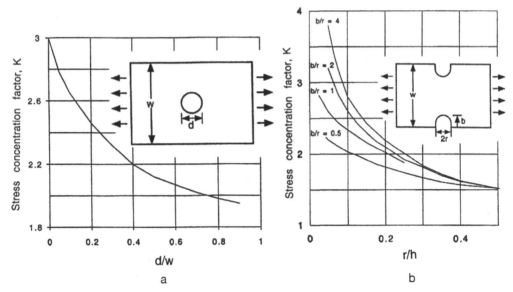

Figure 11.12. Theoretical stress concentration factors. Adapted from G. Neugebauer, *Production Engineering*, v.14 (1943).

in fatigue strength, $K_f = 1$, $q = 0$. The value of q increases with strength level and notch radius, ρ. Several empirical equations for calculating have been proposed. Neuber* suggested for steels that

$$q = 1/[1 + \sqrt{(\beta/\rho)}], \qquad (11.12)$$

where (in mm) is given by

$$\log\beta = -(\sigma_u - 134\,\text{MPa})/586. \qquad (11.13)$$

Here σ_u is the tensile strength. Figure 11.13 shows values of the notch sensitivity factor, q, calculated from Equations 11.12 and 11.13. The notch sensitivity increases with strength level and decreases with increasing notch sharpness.

EXAMPLE PROBLEM #11.4: Calculate the stress concentration factor for fatigue, K_f, for a plate of steel 2 in. wide and 0.25 in. thick with a hole 0.5 in. diameter in the center. The steel has a tensile strength of 600 MPa.

Solution: $d/w = 0.125$. From Figure 11.12a, $K_t = 2.6$. The notch radius is 0.125 in. = 3.18 mm. From Figure 11.13 for 600 MPa, $q = 0.96$, $K_f = 0.96 \times 2.6 = 2.5$.

---

* H. Neuber, *Theory of Notch Stresses*, J. W. Edwards (1946).

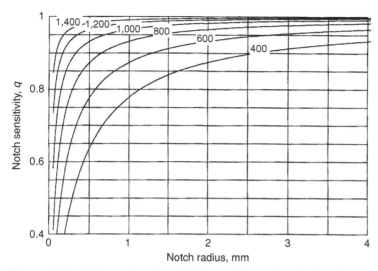

Figure 11.13. Values of notch sensitivity, q, for steels calculated from Equations 11.12 and 11.13. The numbers are the tensile strengths in MPA. Note that q increases with tensile strength and larger notch radii.

Surface Conditions

Fatigue cracks usually start on the surface.* This is because most forms of loading involve some bending or torsion, so the stresses are greatest at the surface. Surface defects also play a role. Therefore, the nature of the surface strongly affects fatigue behavior. There are three important aspects of the surface: hardness, roughness, and residual stresses.

In general, increased surface hardness increases fatigue limits. Carburizing, nitriding, flame, and induction hardening are used to harden surfaces and increase the fatigue strengths. Different finishing operations influence surface topography. Valleys of rough surfaces act as stress concentrators, so fatigue strength decreases with surface roughness. Surfaces produced by machining are generally smoother than cast or forged surfaces. Grinding and polishing further increase smoothness. The use of polished surfaces to improve fatigue behavior is not warranted where exposure to dirt and corrosion during service may deteriorate the polished surface. Figure 11.14 shows the effects of surface condition. The effects of a corrosive environment are also clear. Indeed, laboratory studies have shown considerable improvement in fatigue behavior when the cyclic stressing was done under vacuum instead of dry air.

---

* The most common exception is where the cyclic loading is from contact with a ball or a cylinder. In this case, the highest stress occurs some distance below the surface. This sort of loading occurs in ball and roller bearings.

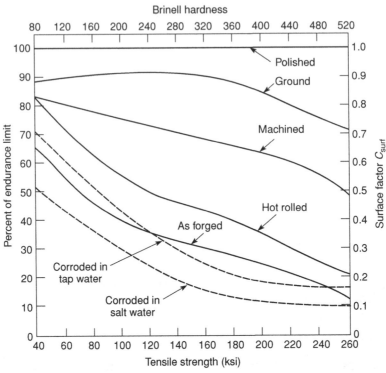

Figure 11.14. Effect of surface finish on fatigue. From C. Lipson and R. C. Juvinall, *Application of Stress, Analysis to Design and Metallurgy*, The University of Michigan Summer Conference, Ann Arbor MI (1961).

Residual stresses play an important role in fatigue. When a part is subjected a load, as in fatigue, the stress at any location is the sum of the residual stress at that point and the stress resulting from the external load (Figure 11.15). Because failures are tensile in nature and start at the surface, residual tension at the surface lowers the resistance to fatigue, whereas residual compression raises the fatigue strength. Note that this effect is in accord with the prediction of the Goodman diagram. Sometimes critical parts are shot peened to produce residual compression in the surface. In this process, the surface is indented with balls that produce local plastic deformation that does not penetrate into the interior of the part. The indentation would expand the surface laterally but the un-deformed interior prevents this leaving the surface under lateral compression.

Some investigators have found that the fatigue strength of a material decreases as the specimen size increases. However, the effect is not large (10% decrease for a diameter increase from 0.1 to 2 in.). The size effect is probably related to the increased amount of surface area where fatigue cracks initiate.

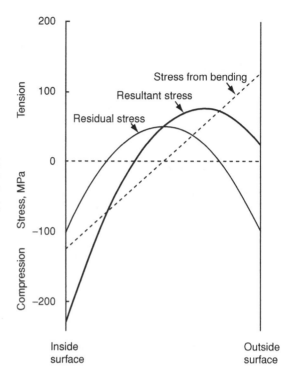

Figure 11.15. Schematic drawing showing the effect of residual stresses. With residual compression in the surface, a larger tensile stress can be applied by bending before there is tension in the surface.

Design Estimates

Shigley* suggested that the endurance limit can be estimated by taking into account the various factors as

$$\sigma_e = \sigma_{eb} C_s C_d (1 - \sigma_m / UTS) / K_f, \qquad (11.14)$$

where σ_{eb} is the base endurance limit (polished un-notched specimen of small diameter cycled about a mean stress of zero), C_s is the correction factor for the surface condition (Figure 11.14), and C_d is the correction factor for specimen size ($C_d = 1$ for $d < 7.6$ mm and 0.85 for $d > 7.6$ mm). The term $(1 - \sigma_m / UTS)$ accounts for the effect of mean stress, σ_m, and $K_f = 1 + q(K_t - 1)$. Equation 11.12 can be used to obtain a first estimate of the endurance limit, but it is always better to use data for the real conditions than to apply corrections to data for another condition.

> **EXAMPLE PROBLEM #11.5:** A round bar of a steel has a yield strength of 40 ksi, a tensile strength of 60 ksi, and an endurance limit of 30 ksi. An elastic analysis indicates that $K_t = 2$. It is estimated that $q = 0.75$. The bar is to be loaded under bending such that a cyclic bending moment of 1500 in-lbs is superimposed on a steady bending moment of 1000 in-lbs.

* J. E. Shigley, *Mechanical Engineering Design*, 3rd ed., McGraw-Hill (1977).

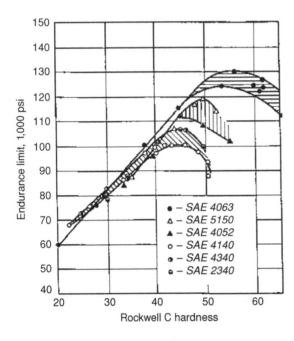

Figure 11.16. Correlation of the endurance limit with hardness for quenched and tempered steels. The endurance generally rises with hardness. From Garwood, Zurburg and Erickson in *Interpretation of Tests and Correlation with Service*, ASM (1951).

The surface has a ground finish. What is the minimum diameter bar that would give an infinite life?

Solution: For a round bar under elastic loading, the stress at the surface is given by $\sigma = Mc/I$, where $c = d/2$ and $I = \pi d^4/64$, so $\sigma = 32M/(\pi d^3)$. $\sigma_m = 10{,}186/d^3$ ksi and $\sigma_a = 15{,}279/d^3$ ksi, where M is the bending moment and d is the bar diameter. $K_f = 1 + q(K_t - 1) = 1.75$. Assuming that $d < 7.6$ mm, $C_d = 1$. From Figure 11.14, $C_s = 0.89$, so the endurance limit is estimated as $\sigma_e = C_s C_d \sigma_{eb}(1 - \sigma_m/UTS)/K_f = (0.89)(1)(30)[1 - 10.86/(60d^3)]/1.75 = 15.25(1 - 0.181/d^3)$. Now equating this to the stress amplitude, $15.279/d^3 = 15.25(1 - 0.181/d^3)$. $d^3 = 1.182$, $d = 1.058$ in. Because this is larger than 7.6 mm, we should use $C_d = 0.85$. Recalculation with $C_d = 0.85$ instead of 1 results in $d = 1.11$ in.

Metallurgical Variables

Because fatigue damage occurs by plastic deformation, increasing the yield strength and hardness generally raises the endurance limit. For steels and titanium alloys, there is a rough rule of thumb that the fatigue limit is about half of the ultimate tensile strength. Figure 11.16 shows the correlation of fatigue limits with hardness in steels. For aluminum alloys, the ratio of the fatigue limit at 10^7 cycles to the tensile strength is between 0.25 and 0.35.

Non-metallic inclusions lower fatigue behavior by acting as internal notches. Alignment of inclusions during mechanical working causes an

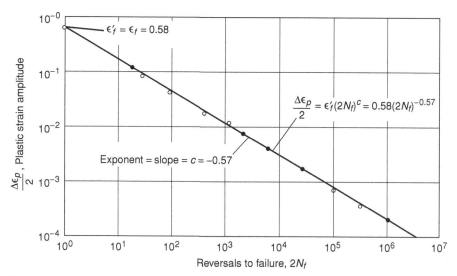

Figure 11.17. A plot of fatigue life, $2N_f$, of an annealed 4340 steel as a function of plastic strain per cycle. The slope is $-c$ and the intercept at one cycle is $\varepsilon_{f'}$. From *Metals Handbook, v. I,* 9th ed., ASM (1978).

anisotropy of fatigue properties. Fatigue strength for loading in the transverse direction (normal to the rolling or extrusion direction) is usually much poorer that for loading in the rolling direction. The number of inclusions can be greatly reduced by vacuum melting or electroslag melting.

Strains to Failure

Cyclic loading in service sometimes subjects materials to imposed forces or stresses. Just as often, materials are subjected to imposed deflections or strains. These are not equivalent if the material strain hardens or strain softens during cycling.

Fatigue could not occur if the deformation during cyclic loading were entirely elastic. Some plastic deformation, albeit very little, must occur during each cycle. This probably accounts for the difference between steel and aluminum alloys. For steels there seems to be a stress below which no plastic deformation occurs. If fatigue data are analyzed by plotting the plastic strain amplitude, $\Delta\varepsilon_p/2$, vs. the number of reversals to failure, $2N_f$, a straight line results. This was first noted by Coffin* and indicates a relationship of the form

$$\Delta\varepsilon_p/2 = \varepsilon_{f'}(2N_f)^{-c}, \tag{11.15}$$

where $\varepsilon_{f'}$ is the true strain at fracture in a tension test ($N = 1$) and $-c$ is the slope. Figure 11.17 is such a plot.

---

* L. F. Coffin, *Trans ASME*, v. 76 (1954).

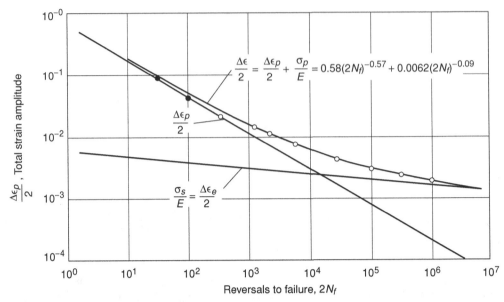

Figure 11.18. The fatigue life, $2N_f$, of an annealed 4340 steel as a function of total strain per cycle. The relation is no longer linear. From *Metals Handbook*, v. I, 9th ed. ASM (1978).

The total strain, including the elastic portion, is $\Delta\varepsilon/2 = \Delta\varepsilon_p/2 + \Delta\varepsilon_e/2$. The elastic term can be expressed as

$$\Delta\varepsilon_e/2 = \sigma_p/E = (B/E)(2N_f)^{-b}, \tag{11.16}$$

where B is a constant that increases with the tensile strength and the exponent, b, is related to the stain hardening exponent. Combining Equations 11.15 and 11.16,

$$\Delta\varepsilon/2 = \varepsilon_{f'}(2N_f)^{-c} + (B/E)(2N_f)^{-b}. \tag{11.17}$$

Figure 11.18 is a plot of the total strain, $\Delta\varepsilon/2$, as well as the elastic and plastic strains vs. $2N_f$. Although $\ln(\Delta\varepsilon_e/2)$ and $\ln(\Delta\varepsilon_p/2)$ vary linearly with $\ln(2N_f)$, $\ln(\Delta\varepsilon/2) = \ln(\Delta\varepsilon_e/2 + \Delta\varepsilon_p/2)$ does not.

The term *low-cycle fatigue* is commonly applied to conditions in which the life, N_f, is less than 10^3 cycles. The term *high-cycle fatigue* is applied to conditions in which the life, N_f, is greater than 10^4 cycles. Note that the range 10^3 to 10^4 cycles corresponds to the range where $\Delta\varepsilon_p > \Delta\varepsilon_e$. For low-cycle fatigue $\Delta\varepsilon_p > \Delta\varepsilon_e$, whereas for high-cycle fatigue $\Delta\varepsilon_e > \Delta\varepsilon_p$. The total life, N_f, can be divided into the period necessary for crack initiation, N_i, and that necessary for the crack to propagate to failure, N_p.

$$N_f = N_i + N_p. \tag{11.18}$$

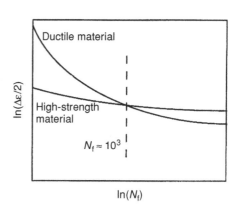

Figure 11.19. Fatigue life as a function of strain amplitude. At large strain amplitudes (low cycle fatigue), the softer, more ductile material will have the longer life while at low strain amplitudes (high cycle fatigue), the stronger, less ductile material will have the longer life.

For low cycle fatigue, crack initiation is rapid and crack propagation accounts for most of the life ($N_p > N_i$), whereas for high cycle fatigue most of the life is spent in crack initiation ($N_i > N_p$).

This leads to the conclusion that under constant strain-amplitude cycling, a ductile material is desirable for low-cycle fatigue and a high-strength material is desirable for high-cycle fatigue. High-strength materials generally have low ductility (low value of $\varepsilon_{f'}$ in Equation 11.17) and ductile materials generally have low strength (low value of B in Equation 11.17). This is shown schematically in Figure 11.19.

Crack Propagation

In the laboratory, crack growth rates may be determined during testing either optically with a microscope or by measuring the electrical resistance. It has been found that the crack growth rate depends on the range of stress intensity factor, ΔK, $\Delta K = K_{max} - K_{min} = \sigma_{max}\sqrt{(\pi a)} - \sigma_{min}\sqrt{(\pi a)} = (\sigma_{max} - \sigma_{min})f\sqrt{(\pi a)}$, or more simply,

$$\Delta K = f\Delta\sigma\sqrt{(\pi a)}, \tag{11.19}$$

where $\Delta\sigma = (\sigma_{max} - \sigma_{min})$.

Figure 11.20 shows schematically the variation of da/dN with ΔK. Below a threshold, ΔK_{th}, cracks do not grow. There are three regimes of crack growth. One is a region where da/dN increases rapidly with ΔK. Then there is large region where $\log(da/dN)$ is proportional to $\log(\Delta K)$, and finally at large values of ΔK_I the crack growth rate accelerates more rapidly. Failure occurs when $K_I = K_{Ic}$. In viewing this figure, one should remember ΔK_I increases with the growth of the crack, so that under constant load a crack will progress up the curve to the right.

In stage II,

$$da/dN = C(\Delta K_I)^m, \tag{11.20}$$

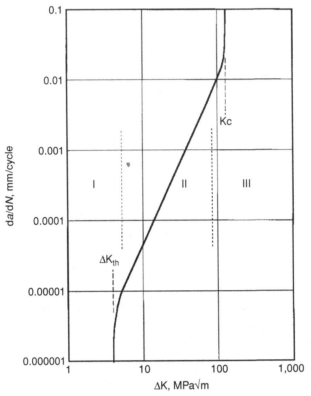

Figure 11.20. A schematic plot of the dependence of crack growth rate, da/dN, on the stress intensity factor, ΔK_I. Because ΔK_I increases with crack length, a, crack growth will progress up the curve. Below the threshold stress, cracks do not grow. In general there are three regions: An initial period (region I), a linear region (II) in which the rate of crack growth is given by Equation 11.19, and a final region (III) of acceleration to failure. The slope of the linear region on the log-log plot equals m.

where C is a material constant that increases with the stress ratio, $R = \sigma_{min}/\sigma_{max}$ (i.e. with the mean stress). The exponent, m, is also material-dependent and is usually in the range of 2 to 7. Equation 11.20 is often called the Paris[*] law.

Note that the crack length at any stage can be found by substituting $(\Delta K_I)^m = [f \Delta \sigma \sqrt{(\pi a)}]^m$ into Equation 11.20 and rearranging,

$$da/a^{m/2} = C(f \Delta \sigma \sqrt{\pi})^m dN. \qquad (11.21)$$

Integration, neglecting the dependence of f on a, gives

$$a^{(1-m/2)} - a_o^{(1-m/2)} = (1 - m/2)C(f \Delta \sigma \sqrt{\pi})^m N. \qquad (11.22)$$

[*] P. Paris and F. Erdogan, *Trans. ASME, J. Basic Engr.* v. D 85 (1963).

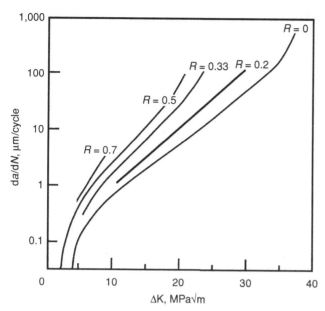

Figure 11.21. Crack growth rate in aluminum alloy 7076-T6 as a function of ΔK for several levels of R. Note that da/dN increases with R. Data from C. A. Martin, NASA TN D5390 (1969).

The number of cycles to reach a crack size a is then

$$N = \left[a^{(1-m/2)} - a_o^{(1-m/2)}\right] / \left[(1 - m/2)C(f\Delta\sigma\sqrt{\pi})^m\right]. \qquad (11.23)$$

The constant, C, in Equations 11.20 through 11.23, depends on the stress ratio, $R = (\sigma_m - \Delta\sigma/2)/(\sigma_m + \Delta\sigma/2)$, as shown in Figure 11.21. This dependence is in accord with the effect of σ_m, predicted by the Goodman diagram. Figure 11.22 shows how a crack length, a, depends on N for two different values of the initial crack length, a_o, and two different values of $\Delta\sigma$.

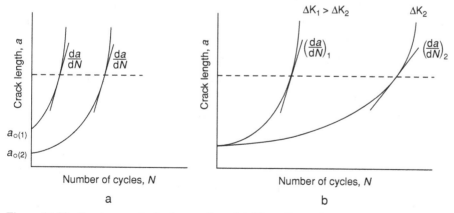

Figure 11.22. Crack growth during cycling. (a) The effect of the initial crack length with constant ΔK. Note that da/dN is the same at the same current crack length, a. (b) The effect of ΔK on crack growth. Note that da/dN increases with ΔK.

EXAMPLE PROBLEM #11.6: Find the exponent, m, and the constant C in Equation 11.20 for aluminum alloy 7076-T6 using the data for $R = 0$ in Figure 11.21.

Solution: $m = \ln[(da/dN)_2/[(da/dN)_1]/[\ln(\Delta K_2/(\Delta K_1)]$. Substituting $(da/dN)_1 = 7 \times 10^{-7}\,m$ at $\Delta K_1 = 10\,\text{MPa}\,\sqrt{m}$ and $(da/dN)_2 = 6 \times 10^{-5}\,m$ at $\Delta K_1 = 30\text{MPa}\sqrt{m}$, $m = \ln 85.7/\ln 3 = 4.05$. $C = (da/dN)_1/(\Delta K_1)^m = 7 \times 10^{-7}/10^{4.05} = 6.2 \times 10^{-11}$.

For low-cycle fatigue, the life can be found by assuming the initial crack size and knowing $\Delta\sigma$, which gives an initial value of ΔK. Integration under the da/dN vs. ΔK curve allows life prediction.

EXAMPLE PROBLEM #11.7: Consider crack growth in 7076-T6 aluminum.

a. Determine the crack growth rate, da/dN, if $\Delta\sigma = 200$ MPa and the crack is initially 1 mm in length. Assume $R = 0$ and $f = 1$.
b. Calculate the number of cycles needed for the crack to grow from 1 mm to 10 mm if $\Delta\sigma = 200$ MPa.

Solution:

a. $\Delta K = f\Delta\sigma\sqrt{(\pi a)} = 2\,\text{MPa}\sqrt{(0.001\pi)} = 11.2\,\text{MPa}\sqrt{m}$. From Figure 11.21, $da/dN \approx 10^{-3}$ mm.
b. Using Equation 11.22 with $m = 4.05$ and $C = 6.2 \times 10^{-11}$ from example Problem #11.6, $N = [(10^{-5})^{-1.025} - (10^{-2})^{-1.025}]/[(-1.0250)(6.2 \times 10^{-8})(200\sqrt{\pi})^{4.05}] = 99$ cycles.

Cyclic Stress-Strain Behavior

Materials subjected to cyclic loading in the plastic range may undergo strain hardening or strain softening in the case of heavily cold-worked material. If the cycling is done at constant strain amplitude, strain hardening causes an increase of stress amplitude during testing. On the other hand, for a material that strain softens, $\Delta\sigma$ decreases during cycling. Figure 11.23 illustrates both of these possibilities for constant plastic-strain amplitude cycling. Cyclic strain hardening can be measured by cycling at one strain amplitude until the stress level saturates and then repeating this at increasing strain amplitudes.

For a material that work-hardens, the strain amplitude will decrease during constant stress amplitude cycling. On the other hand, for material that work-softens, the strain amplitude will decrease during constant stress amplitude cycling. In contrast, if a material is cycled under constant stain amplitude, the stresses will increase in a material that work-hardens and decrease in a material that work-softens. These behaviors are illustrated in Figures 11.24a and b.

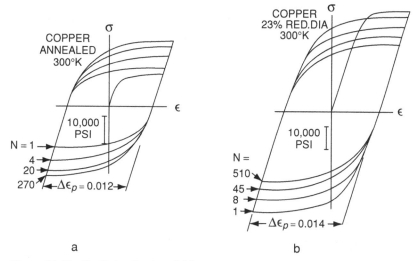

a   b

Figure 11.23. Cyclic hardening of (a) annealed copper and cyclic softening of a (b) cold-worked copper tested under constant plastic strain cycling. From C. E. Feltner and C. Laird, *Acta Met*, v. 15 (1967).

Temperature and Cycling Rate Effects

Increased temperatures lower flow stresses, and therefore there is more plastic flow per cycle in constant stress amplitude cycling. Near room temperature, this effect is more pronounced in polymers than in metals. With higher frequencies, there is less plastic deformation per cycle so the crack growth rate is lower. On the other hand, with high cycling rates the material's temperature may rise unless the heat generated by the mechanical hysteresis is dissipated. This is much more important with polymers than with metals because

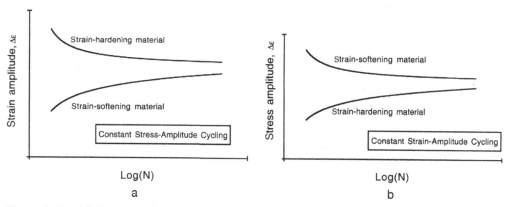

a   b

Figure 11.24. (a) Change of strain amplitude during constant stress amplitude cycling. The strain amplitude should decrease for a strain-hardening material and increase for a strain softening material. (b) Change of stress range during constant strain amplitude cycling. The stress amplitude should increase for a strain-hardening material and decrease for a strain-softening material.

they tend to have larger hysteresis losses and lower thermal conductivities. With higher temperatures, there is more plastic strain per cycle under constant stress cycling, so cracks grow more rapidly and the life will decrease. Thus, increased frequency may either shorten or lengthen fatigue life. The net effect of frequency for constant stress amplitude, if any, will depend on which effect predominates.

The effects of frequency at constant strain amplitude cycling are different. Heating of the material and lower frequencies both cause lower stress amplitudes. However, the cumulative strain to failure should remain unaffected. Unless the mean stress is zero ($R = 0$), creep during cyclic loading will cause the material to permanently elongate because the stress on the tensile half-cycle will be larger than on the compressive half-cycle.

Thermal effects are much more important in polymers than in metals. For most polymers, room temperature is about half the absolute melting point, T_M, and thermal softening becomes important at $(0.4$ to $0.5)T_M$. Likewise, the effect of cyclic frequency is more important.

Fatigue Testing

Fatigue tests are often made using round bars rotated under a bending load so that the entire surface is subjected to alternating tension and compression. Flat specimens from sheet or plate may be tested in fatigue by bending. In either case, the testing machine may be made to impose a constant deflection or a constant bending force. Fatigue testing with axial loading is normally much slower and requires great care. In this case, tension-release is often used instead of tension-compression.

Design Considerations

There are two methods of designing to prevent fatigue failure: designing for an infinite life without inspection or designing for periodic inspection. For many parts, inspection is not a possibility. An automobile axle is an example. It must be designed with enough of a safety factor so that it will not fail within the life of the car. In these cases, it is usual to assume initiation-controlled fatigue life where endurance limit and life are improved by increased hardness.

The other possibility is to design for periodic inspection. This is done in the aircraft industry where safety factors must be lower to minimize weight. The interval between inspections must be short enough in terms of service so that a crack of the largest undetectable length cannot grow to failure before the next inspection. In this case, the design and the inspection schedule are based on crack growth rate, da/dN. Fatigue cracks can be arrested by drilling holes at their tips. This requires the crack to re-initiate.

Notes

Railroad accidents in the middle of the nineteenth century stimulated the first research on the nature of fatigue. Poncelet first used the term "fatigue" in lectures as early as 1839. Inspection of railroad cars for fatigue cracks was recommended in France as early as 1853. About 1850, a number of papers on fatigue were written in England. A. Wohler (1819–1914), a German railroad engineer, did the first systematic research on fatigue. He recognized the effects of sharp corners. He also noted that the maximum stress allowable depended on the stress range, and that the maximum allowable stress increased as the minimum stress increased.

Built by de Haviland the Comet was the first commercial passenger jet airliner. It was put into service in May 1952 after years of design, production, and testing. Almost two years later, in January and May 1954, two Comets crashed into the Mediterranean near Naples after taking off from Rome. The cause of the failures was discovered only after parts of one plane that crashed into the Mediterranean were retrieved by the British navy from a depth of over 300 ft. It was found that fatigue was responsible. Cracks originated from rivet holes near the windows. This conclusion was confirmed by tests on a fuselage that was repeatedly pressurized and depressurized. The sharp radii of curvature at the corners acted as stress concentrators. Apparently, the engineers had overlooked the stresses to the fuselage caused by cycles of pressurization and depressurization during take-off and landing. Up to this time, the main fatigue concern of aircraft designers was with engine parts that experience a very large number of stress cycles, and low cycle fatigue of the fuselage had not been seriously considered. As a result of the investigations, the Comets were modified and returned to service a year later.

Resistance to fatigue can be improved by autofretage. The idea is that if a part is overstressed once in tension so that a small amount of plastic deformation occurs, it will be left in residual compression. In the past, cast bronze cannons were over-pressurized so that the internal surface of the bore would be left in compression. It has been suggested (with tongue in cheek) that the underside of aircraft wings could be overstressed by having the plane sharply pull out of a steep dive, causing the wing tips to bend upward.

REFERENCES

S. Suresh, *Fatigue of Materials*, Cambridge U. Press (1991).

L. F. Coffin Jr., *Trans ASME*, v76 (1954).

P. C. Paris, *Fatigue- An Interdisciplinary Approach, Proc. 10th Sagamore Conf.*, Syracuse U. Press (1964).

R. W. Hertzberg, *Deformation and fracture of Engineering Materials*, 4th Ed., Wiley (1995).

T. H. Courtney, *Mechanical Behavior of Materials*, 2nd ed. McGraw Hill (2000).

J. E. Shigley, *Mechanical Engineering Design*, 3rd Ed., McGraw-Hill (1977).

R. O. Ritchie *Mater. Sci & Engrg.*, v. 103 (1988).

R. W. Hertzberg & J. A. Manson, *Fatigue of Engineering Plastics*, Academic Press (1980).

Problems

1. For steels, the endurance limit is approximately half the tensile strength, and the fatigue strength at 10^3 cycles is approximately 90% of the tensile strength. The *S-N* curves can be approximated by a straight lines between 10^3 and 10^6 cycles when plotted as $\log(S)$ vs. $\log(N)$. Beyond 10^6 cycles, the curves are horizontal.
   A. Write a mathematical expression for *S* as a function of *N* for the sloping part of the *S-N* curve, evaluating the constants in terms of the approximations given.
   B. A steel part fails in 12,000 cycles. Use the given expression to find what percent decrease of applied (cyclic) stress would be necessary to increase in the life of the part by a factor of 2.5 (to 30,000 cycles).
   C. Alternatively, what percent increase in tensile strength would achieve the same increase in life without decreasing the stress?

2. A. Derive an expression relating the stress ratio, *R*, to the ratio of cyclic stress amplitude, σ_a, to the mean stress, σ_m.
   B. For $\sigma_a = 100$ MPa, plot *R* as a function of σ_m over the range $0 \leq \sigma_m \leq 100$ MPa.

3. The properties of a certain steel are given in Table 11.1.
   A. Plot a modified Goodman diagram for this steel showing the lines for yielding as well as fatigue failure.
   B. For each of the cyclic loadings given, determine whether yielding, infinite fatigue life, or finite fatigue life is expected.
      i. $\sigma_{max} = 250$ MPa, $\sigma_{min} = 0$ ii. $\sigma_{max} = 280$ MPa, $\sigma_{min} = -200$ MPa
      iii. $\sigma_{mean} = 280$ MPa, $\sigma_a = -70$ MPa iv. $\sigma_{mean} = -70$ MPa, $\sigma_a = 140$ MPa

4. The notch sensitivity factor, *q*, for gray cast iron is very low. Offer an explanation in terms of the microstructure.

5. A 1040 steel has been heat-treated to a yield stress of 900 MPa and a tensile yield strength of 1330 MPa. The endurance limit (at 10^6 cycles for cyclical loading about a zero mean stress) is quoted as 620 MPa. Your

Table 11.1. *Properties of a steel*

Tensile strength	460 MPa
Yield strength	300 MPa
Endurance limit	230 MPa

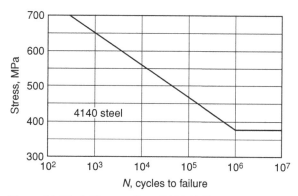

Figure 11.25. *S-N* curve for a SAE 4140 steel.

company is considering using this steel with the same heat treatment for an application in which fatigue may occur during cyclic loading with $R = 0$. Your boss is considering shot peening the steel to induce residual compressive stresses in the surface. Can the endurance limit be raised this way? If so, by how much? Discuss this problem with reference to the Goodman diagram using any relevant calculations.

6. Low cycle fatigue was the cause of the Comet failures. Estimate how many pressurization-depressurization cycles the planes may have experienced in the two years of operations. An exact answer is not possible, but by making a reasonable guess of the number of landings per day and the number of days of service, a rough estimate is possible.

7. Frequently, the *S-N* curves for steel can be approximated by a straight line between $N = 10^2$ and $N = 10^6$ cycles when the data are plotted on a log-log scale, as shown in the figure for SAE 4140 steel. This implies $S = AN^b$, where A and b are constants.

   A. Find b for the 4140 steel for a certain part made from 4140 steel (Figure 11.25).

   B. Fatigue failures occur after five years. By what factor would the cyclic stress amplitude have to be reduced to increase the life to ten years? Assume the number of cycles of importance is proportional to the time of service.

8. Figure 11.26 shows the crack growth rate in aluminum alloy 7075-T6 as a function of ΔK for $R = 0$. Find the values of the constants C and m in Equation 12.19 that describe the straight-line portion of the data. Give units.

9. Find the number of cycles required for a crack to grow from 1 mm to 1 cm in 7075-T6 (Problem 8) if $f = 1$ and $\Delta \sigma = 10$ MPa. Remember that $\Delta K = f \Delta \sigma \sqrt{(\pi a)}$.

10. The fatigue life of a certain steel was found to be 10^4 cycles at ± 70 MPa was and 10^5 cycles at ± 50 MPa. A part made from this steel was given 10^4

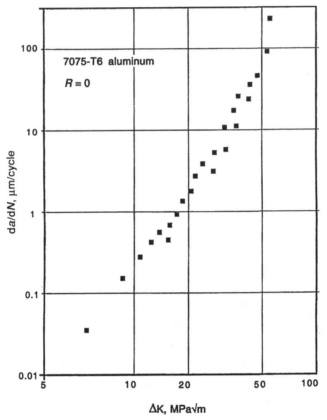

Figure 11.26. Crack growth rate of 7075-T6 aluminum for $\Delta K = 0$. Data from C. M. Hudson, *NASA TN D-5300* (1969).

cycles at ± 61 MPa. If the part were then cycled at ± 54 MPa, what would be its expected life?

11. For another steel, the fatigue limits are 10,000 cycles at 100 ksi, 50,000 cycles at 75 ksi, and 200,000 cycles at 62 ksi. If a component of this steel had been subjected to 5000 cycles at 100 ksi and 10,000 cycles at 75 ksi, how many additional cycles at 62 ksi would cause failure?

12. In fatigue tests on a certain steel, the endurance limit was found to be 1000 MPa for $R = 0$ (tensile-release) ($\sigma_a = \sigma_m = 500$ MPa) and 1000 MPa for $R = -1$ (fully reversed cycling). Calculate whether a bar of this steel would fail by fatigue if it were subjected to a steady stress of 600 MPa and a cyclic stress of 500 MPa.

12 Polymers and Ceramics

Introduction

Up to this point, the treatment has emphasized metallic materials because metals are most widely used for their mechanical properties. This chapter covers the differences between the properties of polymers and ceramics on the one hand and metals on the other.

Elasticity of Polymers

Elastic moduli of thermoplastic polymers are much lower and much more temperature sensitive than those of metals. Figure 12.1 illustrates schematically the temperature dependence of the elastic moduli of several types of polymers. The temperature dependence is greatest near the glass-transition temperature and near the melting point. The crosslinked polymer cannot melt without breaking the covalent bonds in the crosslinks. The stiffness of a polymer at room temperature depends on whether its glass-transition temperature is above or below the room temperature. Below the glass-transition temperature, the elastic moduli are much higher than above it. Figure 12.2 indicates that the modulus of polystyrene changes by a factor of more than 10^3 between 85°C and 115°C.

Glass Transition

If a random linear polymer is cooled very slowly, it may crystallize. Otherwise, it will transform to a rigid glass at its *glass-transition temperature*, T_g. Figure 12.3 is a plot of how the volume may change. If it crystallizes, there is an abrupt volume change. If it does not crystallize, there is a change of slope at T_g. Other properties change as the polymer is cooled below T_g. It toughness and ductility sharply decrease and its Young's modulus greatly increases. Below T_g, a polymer is in a glassy state. Its molecules are virtually frozen in place.

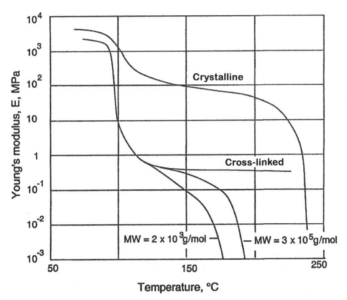

Figure 12.1. Temperature dependence of E for several types of polymers. Crystallization, crosslinking, and increased molecular weight increase stiffness at higher temperatures. Data from A. V. Tobolsky, *J. Polymer Sci.*, Part C, Polymer Symposia No. 9 (1965).

The glass transition is a second-order transition: There is no transfer of heat, but the heat capacity does change. The volume changes to accommodate the increased motion of the wiggling chains, but it does not change discontinuously.

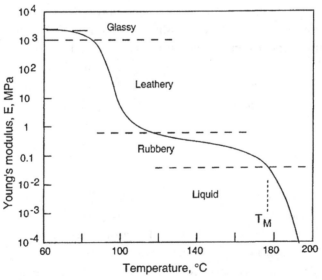

Figure 12.2. Temperature dependence of Young's modulus of polystyrene. As the temperature is increased form 80° to 120°C through the glass transition, the modulus drops by more than a factor of three. Data from A. V. Tobolsky, *ibid.*

Table 12.1. *Glass transition temperatures of several common polymers*

Polymer	T_g (°C)
Polyethylene (LDPE)	−125
Polypropylene (atactic)	−20
Poly(vinyl acetate) (PVAc)	28
Poly(ethyleneterephthalate) (PET)	69
Poly(vinyl alcohol) (PVA)	85
Poly(vinyl chloride) (PVC)	81
Polypropylene (isotactic)	100
Polystyrene	100
Poly(methylmethacrylate) (atactic)	105

The glass phase is not at equilibrium, so the value of T_g is not unique. It depends on the molecular weight and cooling rate. Approximate glass transition temperatures of a few polymers are listed in Table 12.1.

Time Dependence of Properties

A strong time (and therefore strain-rate) dependence of the elastic modulus accompanies the large temperature dependence, as shown in Figure 12.4. This is because the polymers undergo time-dependent viscoelastic deformation when stressed. If a stress is suddenly applied, there is an immediate elastic response. More deformation occurs with increasing time. The effect is so large near the glass-transition temperature that it is customary to define the modulus in terms of the time of loading.

Figure 12.5 illustrates the time dependence of Young's modulus. Under constant load the strain increases with time, so $E = s/e$ decreases with time. Figure 12.4 suggests that the time-dependence of Young's modulus above T_g

Figure 12.3. Changes in specific volume as a random linear polymer is cooled.

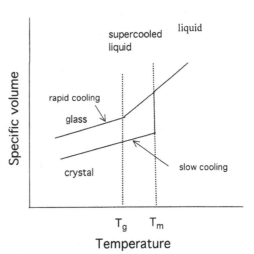

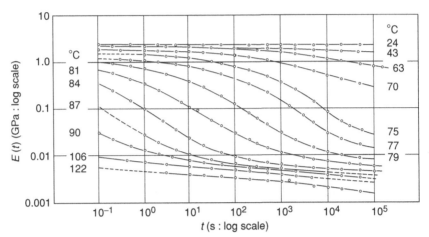

Figure 12.4. Time-dependence of Young's modulus of PVC between 24°C and 122°C. From N. G. McCrum, C. P. Buckley, and C. B. Bucknall, *Principles of Polymer Engineering*, Oxford (1988).

depends on the temperature, and that the dependence at one temperature is related to that at another temperature by a translation along the time scale. Williams, Landel, and Ferry* suggested that the translation, a_T, is given by

$$\log(a_T) = C_1(T - T_g)/(C_2 + T - T_g) \tag{12.1}$$

where C_1 and C_2 are constants. It was suggested that $C_1 = 14.7$ and $C_2 = 51.6\,\mathrm{K}$ for most polymers. This equation (which is known as the WLF equation) can be used to predict long-time behavior from shorter-time tests at a higher temperature.

Rubber Elasticity

The elastic behavior of rubber is very different from that of crystalline materials. Rubber is a flexible polymer in which the molecular chains are crosslinked. The number of crosslinks is controlled by the amount of crosslinking agent compounded with the rubber. Elastic extension occurs by straightening of the chain segments between the crosslinks. With more crosslinking, the chain segments are shorter and have less freedom of motion, so the rubber is stiffer. Originally, only sulfur was used for crosslinking, but today other crosslinking agents are used. Under a tensile stress, the end-to-end lengths of the segments increase but thermal vibrations of the free segments tend to pull the ends together, just as the tension in violin strings is increased by their vibration. A mathematical model that describes the elastic response under uniaxial

* M. L. Williams, R. F. Landel and J. D. Ferry, *J. Amer. Chem Soc.* v77 (1955) p. 3701.

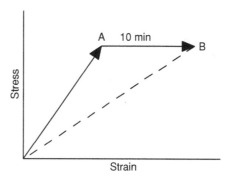

Figure 12.5. Schematic drawing showing why E is time-dependent. After loading, the specimen undergoes an instantaneous elastic strain to point A. After 10 minutes under constant stress, the specimen undergoes additional viscoelastic (anelastic) strain to point B. For the same stress, the strain is greater so the modulus is lower.

tension or compression predicts

$$\sigma = G(\lambda - 1/\lambda^2), \tag{12.2}$$

where λ is the extension ratio, $\lambda = L/L_o = 1 + e$, and e is the engineering strain. The shear modulus, G, depends on the temperature and on the length of the free segments,

$$G = NkT, \tag{12.3}$$

where N is the number of chain segments/volume and k is Boltzmann's constant. Figure 12.6 shows the theoretical predictions of the equation with σ normalized by G. Agreement with experiment is very good for $\lambda < 1.5$, but is poorer at larger values of λ.

Yielding

Figure 12.7 shows typical tensile stress-strain curves for a thermoplastic. The lower strengths at higher temperatures are obvious. At low temperatures (e.g., $-25°C$), PMMA is brittle. At higher temperatures, the initial elastic region is followed by a drop in load that accompanies yielding. A strained region or neck forms and propagates the length of the tensile specimen (Figures 12.8 and 12.9). Only after the whole gauge section is strained does the stress again

Figure 12.6. Stress–extension (λ) curve of vulcanized natural rubber with $G = 0.39\,\text{MPa}$. The continuous curve is the prediction of Equation 12.1. The dashed curve is from data of L. R.G. Treloar, *The Physics of Rubber Elasticity, 3rd Ed.* Oxford (1975). Theory and experiment agree closely for $\lambda < 1.5$.

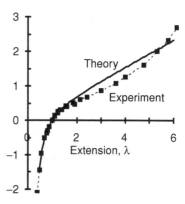

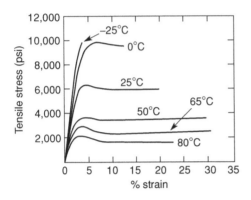

Figure 12.7. Tensile stress-strain curves for poly-methyl methacrylate (PMMA) at several temper-atures. Note that the strength decreases and the elongation increases with higher temperatures. From Carswell and H. K. Nason, *ASTM Symposium on Plastics*, Philadelphia (1944).

rise. Superficially, this is similar to the upper and lower yield points and propagation of Lüder's bands in low-carbon steel after strain aging. However, there are several notable differences.

The engineering strain in the necked region of a polymer is several hundred percent, in contrast to 1% or 2% in a Lüder's band in steel. Also, the mechanism is different. The yield point of steels is associated with the segregation of carbon to dislocations. In polymers, the necking deformation is associated with the reorientation of the polymer chains in the deformed material, so that after the deformation the chains are aligned with the extension axis (see Figure 12.10.) This causes a very large increase in the elastic modulus. Continued stretching after the necked region has propagated the length

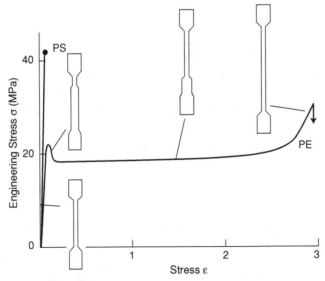

Figure 12.8. Stages of neck formation in a linear polymer. The deformation is uniform up to the maximum load, then it is localized into a band that gradually spreads through the gauge section. After it consumes the whole gauge section, all the molecules are aligned with the tensile axis and the load rises sharply. Note the much larger elastic modulus at the end.

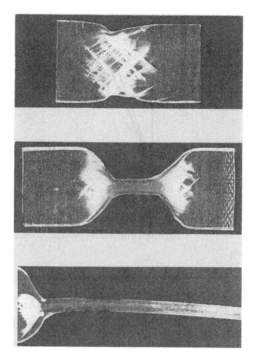

Figure 12.9. Stress-strain curve for polyethylene. Note the formation and propagation of a necked region along the gauge section.

of the gauge section can be achieved only by continued elastic deformation. This now involves the opening of the bond angle of the C-C bonds. The stress rises rapidly until the specimen fractures. If the specimen is unloaded before fracture and a stress is applied perpendicular to the axis of prior extension, the material will fail at very low loads because only van der Waals bonds need be broken.

Although large molecular weights tend to result in stronger and more ductile polymers, they also raise the viscosity of the molten polymer, which is often undesirable in injection molding.

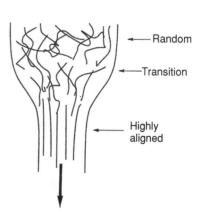

Figure 12.10. Schematic illustration of alignment of molecules during neck formation.

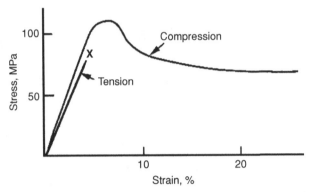

Figure 12.11. Stress-strain curves of an epoxy measured in tension and compression tests. Data from P. A. Young and R. J. Lovell, *Introduction to Polymers*, *2nd Ed.* Chapman and Hall (1991).

Effect of Pressure

For polymers, the stress-strain curves in compression and tension can be quite different. Figure 12.11 is the stress strain curve for epoxy in tension and compression. The stress-strain curve in compression is considerably higher.

Crazing

Many thermoplastics undergo a phenomenon known as *crazing* when loaded in tension. Sometimes the term *craze yielding* is used to distinguish this from the usual *shear yielding*. A craze is an opening resembling a crack. However, a craze is not a crack in the usual sense. Voids form and elongate in the direction of extension. As a craze advances, fibers span the opening, linking the two halves. Figure 12.12 is a schematic drawing of a craze, and Figure 12.13 shows crazes in polystyrene.

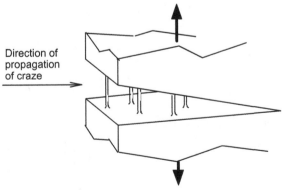

Figure 12.12. Schematic drawing of a craze. Note the fibrils connecting both sides of the craze.

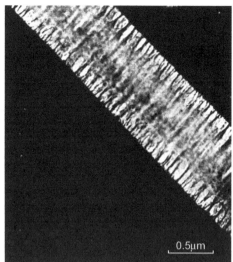

Figure 12.13. Typical microstructure of a thin craze in polystyrene. Note the fibrils that span the crack. From P. Behan, M. Bevis and D. Hull, *Phil. Mag.*, v. 24 (1971).

Fracture

Polymers like most metals are embrittled at low temperatures. The ductile-brittle transition is associated with the glass-transition temperature. Considerable effort has gone into developing polymers that are both strong and tough. Polycarbonate is useful in this respect. Rubber particles in polymers such as polystyrene add greatly to the toughness with some decrease of strength. The ductility of PMMA and the toughness of epoxy are increased by additions of rubber.

Ceramics

Most ceramics are very hard and have very limited ductility at room temperature. Their tensile strengths are limited by brittle fracture but their compressive strengths are much greater. Ceramics tend to retain high hardnesses at elevated temperatures, so they are useful as refractories such as furnace linings and as tools for high-speed machining of metals. Most refractories are oxides, so unlike refractory metals oxidation at high temperature is not a problem. The high hardness of ceramics at room temperature leads to their use as abrasives, either as loose powder or bonded into grinding tools. The low ductility of ceramics limits the structural use of ceramics mainly to applications in which the loading is primarily compressive.

Weibull Analysis

Data on fracture strength typically have a very large amount of scatter. In Chapter 10, it was shown that the fracture strength, σ_f, of a brittle material is

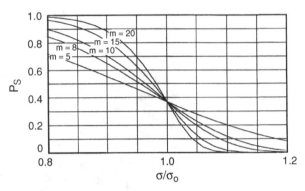

Figure 12.14. Probability of survival at a stress, σ, as a function of σ/σ_o.

proportional to $K_{Ic}/\sqrt{(\pi a)}$, where a is the length of a pre-existing crack. The scatter of fracture strength data is caused mainly by statistical variations in the length of pre-existing cracks. For engineering use, it is important to determine not only the average strength but also the amount of its scatter. If the scatter is small, one can safely apply a stress only slightly below the average strength. On the other hand, if the scatter is large the stress in service must be kept far below the average strength. Weibull* suggested that in a large number of samples, the fracture data could be described by

$$P_s = \exp[-(\sigma/\sigma_o)^m], \tag{12.4}$$

where P_s is the probability that a given sample will survive a stress of σ without failing. The terms σ_o and m are constants characteristic of the material. The constant σ_o is the stress level at which the survival probability is $1/e = 0.368$, or 36.8%. A large value of m in Equation 12.4 indicates very little scatter, and conversely a low value of m corresponds to a large amount of scatter. This is shown in Figure 12.14.

The relative scatter depends on the "modulus," m.

Because

$$\ln(-\ln P_s) = m\ln(\sigma/\sigma_o), \tag{12.5}$$

a plot of P_σ on a log(log) scale vs. σ/σ_o on a log scale is a straight line with the slope is m as shown in Figure 12.15.

Porosity

Most ceramic objects are made from powder by pressing and sintering, so they contain some porosity. Although porosity is desirable in materials used

---

* W. Weibull, *J. Appl. Mech.*, v. 18 1951, p. 293. and *J. Mech. Phys. Solids*, v. 8 (1960).

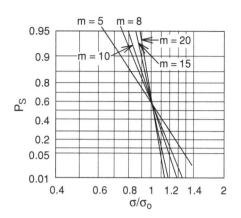

Figure 12.15. Probability of survival at a stress, σ, as a function of σ/σ_o. Note that this is the same as Figure 12.14, except here Ps is plotted on a log(log) scale and σ/σ_o on a log scale. The slopes of the curves are the values of m.

for insulation and filters, porosity is usually undesirable because it adversely affects the mechanical properties. The elastic modulus decreases with porosity. This is shown in Figure 12.16 for the case of alumina. It can be seen that a 10% porosity causes a decrease of the modulus of about 20%. The effects of porosity on strength and on creep rate at elevated temperatures are even more pronounced. The solid line in Figure 12.17 is the prediction of the equation

$$\sigma = \sigma_o \exp(-bP), \qquad (12.6)$$

where P is the volume fraction porosity and the constant, b, is about 7. Fracture toughness also falls precipitously with porosity.

EXAMPLE PROBLEM #12.3: When the porosity of a certain ceramic is reduced from 2.3% to 0.5%, the fracture stress drops by 12%. Assuming Equation 12.6 applies, how much would the fracture stress increase above the level for 0.5% porosity if all porosity were removed?

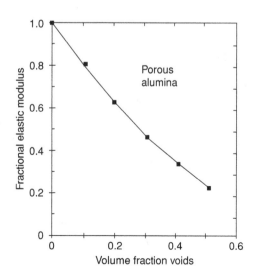

Figure 12.16. Decrease of the elastic modulus with porosity. The fractional modulus is ratio of the modulus of porous alumina to that of 100% dense alumina. Data from R. L. Coble and W. D. Kingery, *J. Amer. Cer. Soc.*, v. 29 (1956).

Table 12.2. *Fracture toughness of some cramics*

Material	K_{Ic} (MPa$\sqrt{m}$)
Al_2O_3 (single crystal)	2.2
Al_2O_3 (polycrystal)	4
mullite (fully dense)	2.0–4.0
ZrO_2 (cubic)	3–3.6
ZrO_2 (partially stabilized)	3–15
MgO	2.5
SiC (hot pressed)	3–6
TiC	3–6
WC	6–20
silica (fused)	0.8
soda-lime glass	0.82
glass ceramics	2.5

Solution: $\sigma_2/\sigma_1 = \exp(-bP_2)/\exp(-bP_1) = \exp[b(P_1 - P_2)]$, $b = \ln(\sigma_2/\sigma_1)/(P_1 - P_2)$. Substituting $\sigma_2/\sigma_1 = 1/0.88$ and $P_1 - P_2 = 0.023 - 0.005 = 0.018$, $b = \ln(1/.88)/0.018 = 7.1$. $\sigma_3/\sigma_2 = \exp[-b(P_3 - P_2)] = \exp[-7.1(0 - 0.005)]) = 1.036$, or a 3.6% increase.

Fracture Toughness

Although all ceramics are brittle, there are significant differences in toughness among ceramics. Typical values of K_{Ic} given in Table 12.2 range from less than 1 to about 7 MPa$\sqrt{m}$. If K_{Ic} is less than about 2, extreme care must be exercised in handling the ceramics. They will break if they fall on the floor. This limits

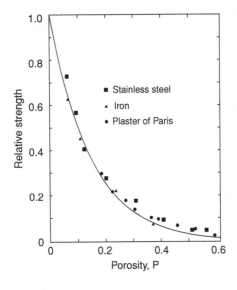

Figure 12.17. Decrease of fracture strength of ceramics and metals with porosity. The solid line represents Equation 12.6 with b = 6. From R. L. Coble and W. D. Kingery, *J. Amer. Cer. Soc.,* v. 29 (1956).

severely the use of ceramics under tensile loading. Ceramics having K_{Ic} greater than about 4 are quite robust. For example, partially stabilized zirconia is used for metal-working tools.

Toughening of Ceramics

Energy absorbing mechanisms can toughen ceramics. There are three basic mechanisms: crack deflection, use of fibers to bridge cracks, and phase trans-formations. Polycrystalline ceramics are usually tougher than single crystals because grain boundaries deflect cracks due to the orientation of the cleav-age planes, which change from grain to grain. Fibers of a second phase bridge across an open crack and continue to carry part of the load. (This is discussed in Chapter 13.) A martensitic transformation that increases volume toughens zirconia that contains CaO or MgO.

Glasses

Glasses are brittle at room temperature. Their fracture toughness can be increased by inducing a residual stress pattern with the surface under compres-sion. This can be done either by a process called *tempering* that cools the sur-face rapidly or by causing large K^+ ions to diffuse into the surface and replace smaller Na^+ ions. Because tempered glass is tougher than untempered glass and breaks into smaller and less dangerous pieces, it is used for side and rear windows of cars. Windshields are made safety glass composed of two pieces of glass laminated with a polymer that keeps the broken shards from causing injury. These forms of glass are illustrated in Figure 12.18.

Thermally Induced Stresses

Like most ceramics, glass is susceptible to stresses caused by thermal gradients. When a material under stress changes temperature, the total stain is given by

$$\varepsilon_x = \alpha \Delta T + (1/E)[\sigma_x - \upsilon(\sigma_y + \sigma_z)]. \tag{12.7}$$

If two regions, A and B, of a piece of material are in intimate contact, they must undergo the same strains. If there is a temperature difference, $\Delta T = T_A - T_B$, between the two regions,

$$\Delta\varepsilon = \alpha_A \Delta T + (1/E)[\sigma_{xA} - \upsilon_A(\sigma_{yA} + \sigma_{zA})]$$
$$= \alpha_B \Delta T + (1/E)[\sigma_{xB} - \upsilon_{AB}(\sigma_{yB} + \sigma_{zB})] \tag{12.8}$$

EXAMPLE PROBLEM #12.4: The temperature of the inside wall of a tube is 200°C and the outside wall temperature is 40°C. Calculate the stresses at

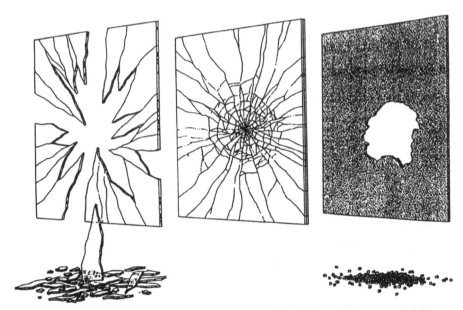

Figure 12.18. Typical fracture patterns of three grades of glass: (a) annealed, (b) laminated, (c) tempered. From *Engineering Materials Handbook IV Ceramics and Glass*, ASM (1991).

the outside of the wall if the tube is made from a glass having a coefficient of thermal expansion of $\alpha = 8 \times 10^{-6}/°C$, an elastic modulus of $10 \times 10_6$ psi, and a Poisson's ratio of 0.3.

Solution: Let x, y, and z be the axial, hoop, and radial directions. The stress normal to the tube wall, $\sigma_z = 0$, and symmetry requires that $\sigma_y = \sigma_x$. Let the reference position be the mid-wall where $T = 120°C$. ΔT at the outside is $40° - 120° = -80°C$. The strains ε_x and ε_y must be zero relative to the mid-wall. Substituting in equation 12.9, $0 = \alpha \Delta T + (1/E)(\sigma_x + \upsilon(\sigma_y + \sigma_z))$, $\alpha \Delta T + (1 - \upsilon)\sigma_x/E = 0$, so $\sigma_x = \alpha E \Delta T/(1-\upsilon)$. $\sigma_x = (8 \times 10^{-6}/°C)(80°C)(10 \times 10^6 \text{psi})/0.7 = 9,140$ psi.

A parameter, R_1, describing the sensitivity of a material to fracturing under thermal shock is given by

$$R_1 = \sigma_f(1 - \upsilon)/(E\alpha), \tag{12.9}$$

where σ_f is the fracture strength of the material. An alternative parameter is

$$R_1 = K_{Ic}(1 - \upsilon)/(E\alpha). \tag{12.10}$$

The elastic properties of most glass are similar, so their resistance to thermal shock is related primarily to their coefficients of thermal expansion, which range from about $10 \times 10^{-6}/°C$ for ordinary bottle or window glass to $3 \times 10^{-6}/°C$ for Pyrex® and $0.5 \times 10^{-6}/°C$ for vycor.

Glassy Metals

Special compositions of alloys have very low liquidus temperatures. Under rapid cooling, these may freeze to an amorphous glass rather than a crystal. For many compositions, the required cooling rates are so great that glass formation is limited to very thin sections. However, several alloys have been developed that form glasses at moderate cooling rates. Relatively large samples of these can be frozen to a glass. Metal glasses have extremely high elastic limits. The elastic strains at yielding may be as large as 2.5%. This permits storage of a very large amount of elastic energy because the elastic energy per volume, E_v, is proportional to the square of the strain.

Notes

One interesting application of metal glass is as plates in the heads of golf clubs. The energy of impact of the club and the ball is absorbed elastically in both the ball and the head of the club. Because there are large losses in the ball, it is advantageous to have as much of the energy as possible absorbed elastically by the head of the club. Glassy metal plates are used to absorb energy as they bend. A large amount of energy can be stored because the metal glass can undergo large elastic strains. There is very little loss of this energy, being almost completely imparted to the ball as it leaves the face of the club.

Before the discovery of vulcanization, natural rubber obtained from the sap of tropical trees became a sticky mass in the heat of summer and hard and brittle in the cold of winter. It was the dream of Charles Goodyear to find a way to make rubber more useful. Without any training in chemistry, he started his experiments in a debtor's prison. In a kitchen of a cottage on the prison grounds, he blended rubber with anything he could find. He found that nitric acid made the rubber less sticky, but when he treated mailbags for the U.S. Post Office and left them for a time in a hot room, they became a sticky mess. In 1839, he accidentally dropped a lump of rubber mixed with sulfur on a hot stove. After scraping the hard mess from the stove, he found that it remained flexible when it cooled. This process of vulcanization was improved and patented a few years later. However, his patent was pirated and he landed in debtor's prison again.

REFERENCES

N. G. McCrum, C. P. Buckley, and C. B. Bucknall, *Principles of Polymer Engineering*, Oxford (1988).

R. J. Young and P. A. Lovell, *Introduction to Polymers*, 2nd Ed., Chapman and Hall (1991).

I. M. Ward and D. W. Hadley, *Mechanical Properties of Solid Polymers*, Wiley (1993).

Engineered Materials Handbook, Vol. 2, Engineering Plastics, ASM International (1988).

S. B. Warner, *Fiber Science*, Prentice Hall (1995).

M. W. Barsoum, *Fundamentals of Ceramics*, McGraw-Hill (1997).

W. D. Kingery, K. Bowen and Uhlman, *Introduction to Ceramics*, 2nd ed. Wiley (1960).

Y.-M. Chiang, D. Birney and W. D. Kingery. *Physical Ceramics*, 2nd ed. Wiley (1997).

Engineered Materials Handbook, vol 4, Ceramics and Glasses, ASM International (1991).

Problems

1. What stress would be required to stretch a piece of PVC by 1% at 75°C (see Figure 12.4). If it were held in the stretched position, what would the stress be after a minute? After an hour? After a day?

2. Evaluate the coefficient of thermal expansion, α, for rubber under an extension of 100% at 20°C. A rubber band was stretched from 6 inches to 12 inches. While being held under a constant stress, it was heated from 20°C to 40°C. What change in length, ΔL, would the heating cause?

3. From a large number of tests on a certain material, it has been learned that 50% of them will break when loaded in tension at stress equal to or less than 520 MPa, and that 30% will break at stress equal to or less than 500 MPa. Assume that the fracture statistics follow a Weibull distribution.
   A. What are the values of σ_o and m in the equation $P_s = \exp[-(\sigma/\sigma_o)^m]$?
   B. What is the maximum permissible stress if the probability of failure is to be kept less than 0.001% (i.e., one failure in 100,000)?

4. Twenty ceramic specimens were tested to fracture. The measured fracture loads in N were 279, 195, 246, 302, 255, 262, 164, 242, 197, 224, 255, 269, 213, 172, 179, 143, 206, 233, 246, and 295.
   a. Determine the Weibull modulus, m.
   b. Find the load for which the probability of survival is 99%.

5. What percent reduction of the elastic modulus of alumina would be caused by 1% porosity? See Figure 12.7.

Introduction

Throughout history, mankind has used composite materials to achieve combinations of properties that could not be achieved with individual materials. The Bible describes mixing of straw with clay to make tougher bricks. Concrete is a composite of cement paste, sand, and gravel. Today, poured concrete is almost always reinforced with steel rods. Other examples of composites include steel-belted tires, asphalt blended with gravel for roads, plywood with alternating directions of fibers, as well as fiberglass-reinforced polyester used for furniture, boats, and sporting goods. Composite materials offer combinations of properties otherwise unavailable. The reinforcing material may be in the form of fibers, particles, or laminated sheets.

Fiber-Reinforced Composites

Fiber composites may also be classified according to the nature of the matrix and the fiber. Examples of a number of possibilities are listed in Table 13.1.

Various geometric arrangements of the fibers are possible. In two-dimensional products, the fibers may be unidirectionally aligned, at 90°to one another in a woven fabric or cross-ply, or randomly oriented (Figure 13.1.) The fibers may be very long or chopped into short segments. In thick objects, short fibers may be random in three dimensions. The most common use of fiber reinforcement is to impart stiffness (increased modulus) or strength to the matrix. Toughness may also be of concern.

Elastic Properties of Fiber-Reinforced Composites

The simplest arrangement is long parallel fibers. The strain parallel to the fibers must be the same in both the matrix and the fiber, $\varepsilon_f = \varepsilon_m = \varepsilon$. For loading parallel to the fibers, the total load, F, is the sum of the forces on the fibers,

Table 13.1. *Various combinations of fibers and matrices*

Fiber	Metal matrix	Ceramic matrix	Polymer matrix
metal	W/Al	concrete/steel	rubber/steel (tires)
ceramic	B/Al	C/glass	fiber glass/polyester
	C/Al	SiC/Si-Al-O-N	fiber glass/epoxy
	Al$_2$O$_3$/Al		
	SiC/Al		
polymer		straw/clay	kevlar/epoxy

F_f, and the matrix, F_m. In terms of the stresses, $F_f = \sigma_f A_f$ and $F_m = \sigma_m A_m$, where σ_f and σ_m are the stresses in the fiber and matrix and where A_f and A_m are the cross-sectional areas of the fiber and matrix. The total force, F, is the sum of $F_f + F_m$,

$$\sigma A = \sigma_f A_f + \sigma_m A_m \tag{13.1}$$

where A is the overall area, $A = A_f + A_m$ and σ is the stress, F/A. For elastic loading, $\varepsilon = E\varepsilon$, $\sigma_f = E_f \varepsilon_f$, and $\sigma_m = E_m \varepsilon_m$, so $E\varepsilon A = E_f \varepsilon_f A_f + E_m \varepsilon_m A_m$. Realizing that $\varepsilon_f = \varepsilon_m = \varepsilon$ and expressing $A_f/A = V_f$ and $A_m/A = V_m$,

$$E = E_f V_f + E_m V_m. \tag{13.2}$$

This is often called the *rule of mixtures*. It is an upper bound to the elastic modulus of a composite.

Consider the behavior of the same composite under tension perpendicular to the fibers. It is no longer reasonable to assume that $\varepsilon_f = \varepsilon_m = \varepsilon$. An alternative, though extreme, assumption is that the stresses in the matrix and the fibers are the same, $\sigma_f = \sigma_m = \sigma$. In this case, $\varepsilon = \sigma/E$, $\varepsilon_f = \sigma_f/E_f$, and $\varepsilon_m = \sigma_m/E_m$ and the overall (average) strain is $\varepsilon = \varepsilon_f V_f + \varepsilon_m V_m$. Combining, $\sigma/E = V_f \sigma_f/E_f + V_m \sigma_m E_m$. Finally realizing that $\sigma_f = \sigma_m = \sigma$,

$$1/E = V_f/E_f + V_m/E_m. \tag{13.3}$$

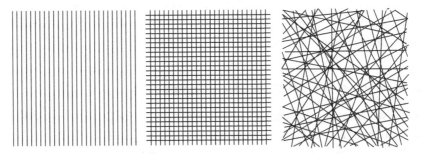

Figure 13.1. Several geometric arrangements of fiber reinforcements.

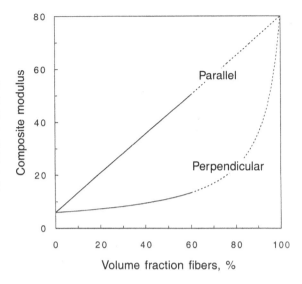

Figure 13.2. Upper and lower bounds to Young's modulus for composites. The upper bound is appropriate for loading parallel to the fibers. Loading perpendicular to the fibers lies between these two extremes. The lines are dashed above $V_f = 60\%$ because this is a practical upper limit to the volume fraction of fibers.

Equation 13.3 is a lower bound for the modulus. Figure 13.2 shows the predictions of equations 13.2 and 13.3. The actual behavior for loading perpendicular to the fibers is between these two extremes.

Now consider the orientation dependence of the elastic modulus of a composite with unidirectionally aligned fibers. Let 1 be the axis parallel to the fibers and let the 2 and 3 axes be perpendicular to the fibers. If a uniaxial stress applied along a direction, x, at an angle, θ, from 1-axis and $90 - \theta$ from 2-axis, the stresses on the 1, 2, 3 axis system may be expressed as

$$
\begin{aligned}
\sigma_1 &= \cos^2\theta \sigma_x, \\
\sigma_2 &= \sigma_3 = \sin^2\theta \sigma_x, \\
\tau_{12} &= \sin\theta\cos\theta \sigma_x. \\
\sigma_3 &= \tau_{23} = \tau_{31} = 0.
\end{aligned} \tag{13.4}
$$

Hooke's laws give the strains along the 1- and 2-axes,

$$
\begin{aligned}
e_1 &= (1/E_1)[\sigma_1 - \upsilon_{12}\sigma_2], \\
e_2 &= (1/E_2)[\sigma_2 - \upsilon_{12}\sigma_1] \\
\gamma_{12} &= (\tau_{12}/G_{12}).
\end{aligned} \tag{13.5}
$$

The strain in the x-direction can be written as

$$
\begin{aligned}
e_x &= e_1\cos^2\theta + e_2\sin^2\theta + 2\gamma_{12}\cos\theta\sin\theta \\
&= (\sigma_x/E_1)[\cos^4\theta - \upsilon_{12}\cos^2\theta\sin^2\theta] + (\sigma_x/E_2)[\sin^4\theta - \upsilon_{12}\cos^2\theta\sin^2\theta] \\
&\quad + (2\sigma_x/G_{12})\cos^2\theta\sin^2\theta.
\end{aligned} \tag{13.6}
$$

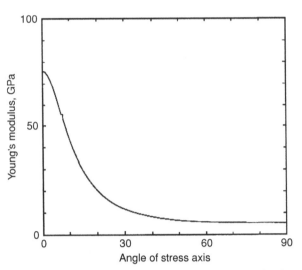

Figure 13.3. The orientation dependence of the elastic modulus in a composite with unidirectionally aligned fibers. Here it is assumed that $E_1 = 75$ MPa, $E_2 = 5$ MPa, $G = 10$ MPa, and $\upsilon_{12} = 0.3$. Note that most of the stiffening is lost if the loading axis is misoriented from the fiber axis by as little as 15°.

The elastic modulus in the x direction, E_x, can be found from Equation 13.6.

$$E_x = \sigma_x/e_x. \tag{13.7}$$

Figure 13.3 shows that the modulus drops rapidly for off-axis loading. The average for all orientations in this figure is about 18% of E_1.

A crude estimate of the effect of a cross-ply can be made from an average of the stiffnesses due to fibers at 0° and 90°, as shown in Figure 13.4. Although the cross-ply stiffens the composite for loading near 90°, it has no effect on the 45° stiffness. Figure 13.5 illustrates why this is so. For loading at 45°, extension can be accommodated by rotation of the fibers without any extension.

The simple averaging used to calculate Figure 13.4 underestimates the stiffness for most angles of loading, θ. It assumes equal strains in both plies in the loading direction but neglects the fact that the strains perpendicular to the loading direction in both plies must also be equal. When this constraint is accounted for, a somewhat larger modulus is predicted. However, this effect disappears at $\theta = 45°$ because composites with both sets of fibers have the same lateral strain.

With randomly oriented fibers, the orientation dependence disappears. One might expect the modulus would be the average of the moduli for all directions of uniaxially aligned fibers. However, this again would be an underestimate because it neglects the fact that the lateral strains must be the same for all fiber alignments. A useful engineering approximation for randomly

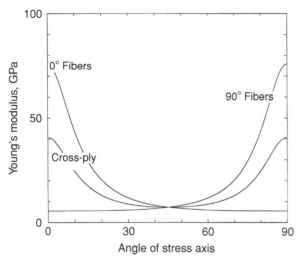

Figure 13.4. Calculated orientation dependence of Young's modulus in composites with singly and biaxially oriented fibers. For this calculation, it was assumed that $E_{\text{para}} = 75$ GPa and $E_{\text{perp}} = 75$ GPa. Note that even with biaxially oriented fibers, the modulus at 45° is very low.

aligned fibers is

$$E \approx (3/8)E_{\text{para}} + (5/8)E_{\text{perp}}, \tag{13.8}$$

where E_{para} and E_{perp} are the E moduli parallel and perpendicular to uniaxially aligned fibers.

Strength of Fiber-Reinforced Composites

The rule of mixtures cannot be used to predict the strengths of composites with uniaxially aligned fibers. The reason can be appreciated by considering the

Figure 13.5. Rotation of fibers in a woven cloth. Stress in at 45° to the fibers allows deformation with little or no stretching of the fibers.

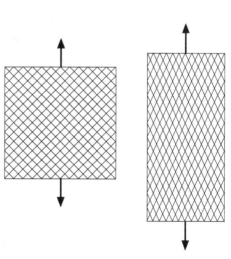

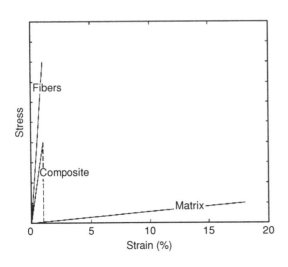

Figure 13.6. Stress-strain curves for the matrix, the fibers, and the composite.

stress-strain behavior of both materials, as shown schematically in Figure 13.6. The strains in the matrix and fibers are equal, so the fibers reach their breaking strengths long before the matrix reaches its tensile strength. Thus, the strength of the composite $UTS < V_m UTS_m + V_f UTS_f$.

If the load carried by the fibers is greater than the breaking load of the matrix, both will fail when the fibers break,

$$UTS = V_m \sigma_m + V_f (UTS)_f, \tag{13.9}$$

where $\sigma_m = (E_m / E_f)(UTS)_f$ is the stress carried by the matrix when the fiber fractures.

For composites with low volume fraction fibers, the fibers may break at a load less than the failure load of the matrix. In this case, after the fibers break the whole load must be carried by the matrix, so the predicted strength is

$$UTS = V_m (UTS)_m. \tag{13.10}$$

Figure 13.7 shows schematically how the strength in tension parallel to the fibers varies with volume fraction fibers. Equation 13.10 applies at low volume fraction fibers and Equation 13.9 applies for higher volume fractions of fibers.

EXAMPLE PROBLEM #13.1: A metal composite composed of metal M reinforced by parallel wires of metal W. The volume fraction W is 50% and W has a yield stress of 200 MPa and a modulus of 210 GPa. Metal M has a yield stress of 50 MPa and a modulus of 30 GPa. What is the maximum stress that the composite can carry without yielding?

Solution: At yielding, the strains will be the same in the wire and matrix. $\sigma_w / E_w = \sigma_m / E_m$. The wire will yield at a strain of $\sigma_w / E_w = 200\,\text{MPa}/210\,\text{GPa} = 0.001$ and the matrix will yield at a strain of

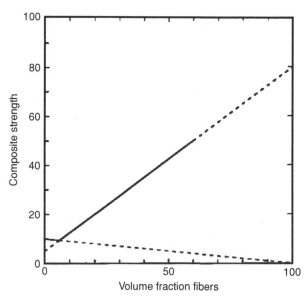

Figure 13.7. Dependence of strength on the volume fraction fibers. It is assumed that the fiber strength, $UTS_f = 80$, $UTS_m = 10$ and $\sigma_m = 5$. Note that the fibers lower the composite strength to less than the tensile strength of the matrix for very low volume fraction fibers.

50 MPa/30 GPa $= 0.0017$. The composite will yield at the lower strain, that is, when $\varepsilon = 0.001$. At this point, the stress in the wire would be 200 MPa and the stress in the matrix would be 0.001×30 GPa $= 30$ MPa. The average stress would be $0.50 \times 200 + 0.50 \times 30 = 115$ MPa.

Volume Fraction of Fibers

Although the stiffness and strength of reinforced composites should increase with the volume fraction of fibers, there are practical limitations on the volume fraction. Fibers must be separated from one another. Fibers are often pre-coated to ensure this separation and to control bonding between fibers and matrix. Techniques of infiltrating fiber arrangements with liquid resins lead to variability in fiber spacing, as shown in Figure 13.8. The maximum possible packing density is greater for unidirectionally aligned fibers than for woven or cross-ply reinforcement. A practical upper limit for volume fraction seems to be about 55% to 60%. This is the reason is why the calculated lines in Figures 13.2 and 13.7 are dashed above $V_f = 60\%$.

Orientation Dependence of Strength

For unidirectionally aligned fibers, the strength varies with orientation. There are three possible modes of failure. For tension parallel or nearly parallel

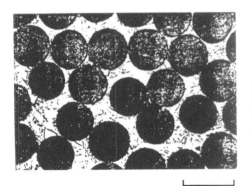

Figure 13.8. Glass fibers in a polyester matrix. Note the variability in fiber spacing. From *Engineered Materials Handbook, v. I, Composites,* ASM International (1987).

0.0127 mm

to the fibers, failure occurs when the stress in the fiber exceeds the fracture strength, S_f, of the fibers. Neglecting the stress in the matrix, the axial stress, σ, at this point is

$$\sigma = S_f/\cos^2\theta. \tag{13.11}$$

For loading at a greater angle to the fibers, failure can occur by shear in the matrix. The axial stress, σ, is

$$\sigma = \tau_{fm}/\sin\theta\cos\theta, \tag{13.12}$$

where τ_{fm} is the shear strength of the fiber-matrix interface. For loading perpendicular or nearly perpendicular to the fibers, failure is governed by the strength, S_{fm}, of the fiber-matrix interface.

$$\sigma = S_{fm}/\sin^2\theta. \tag{13.13}$$

These three possibilities are shown in Figure 13.9. The three fracture modes are treated separately. However, the slight increase of failure stress, σ, with θ for low angles is not realized experimentally. This fact suggests an interaction

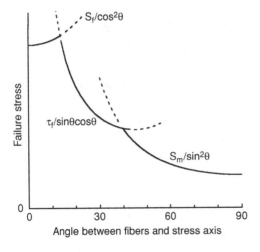

Figure 13.9. Failure strength of a unidirectionally aligned fiber composite as a function of orientation. There are three possible fracture modes: tensile fracture of the fibers, shear failure parallel to the fibers, and tensile failure normal to the fibers.

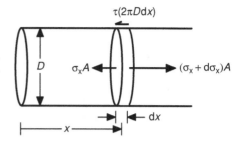

Figure 13.10. Force balance on a differential length of fiber. The difference in the normal forces must be balanced by the shear force.

of the longitudinal and shear fracture modes. Also, for angles near 45°, experimental values of fracture strength tend to fall somewhat below the predictions in Figure 13.9, which indicates an interaction between shear and transverse fracture modes.

Fiber Length

Fabrication is much simplified if the reinforcement is in the form of chopped fibers. Chopped fibers can be blown onto a surface to form a mat. Composites with chopped fibers can be fabricated by processes that are impossible with continuous fibers, such as extrusion, injection molding, and transfer molding. The disadvantage of chopped fibers is that some of the reinforcing effect of the fibers is sacrificed because the average axial stress carried by fibers is less for short ones than for long ones. The reason is that at the end of the fiber, the stress carried by the fiber is vanishingly low. Stress is transferred from the matrix to the fibers primarily by shear stresses at their interfaces. The average axial stress in a fiber depends on its aspect ratio, D/L, where D and L are the fiber's diameter and length.

The following development is based on the assumption that the shear stress between the matrix and the fiber, τ, is constant and that no load is transferred across the end of the fiber. Figure 13.10 shows a force balance on a differential length of fiber, dx, which results in $(\sigma_x + d\sigma_x)A = \sigma_x A + \tau(\pi D dx)$ or $(D^2/4)d\sigma_x = \tau(D dx)$. Integrating,

$$\sigma_x = 4\tau(x/D). \tag{13.14}$$

This solution is valid only for $\sigma_x \leq \sigma_\infty$ where σ_∞ is the stress that would be carried by an infinitely long fiber. Figure 13.11 illustrates three possible conditions depending on x^*/L, where $x^* = (D/4)(\sigma_\infty/\tau)$ is the distance at which $\sigma_x = \sigma_\infty$. The length, $L^* = 2x^*$ is called the *critical length*.

If $L > L^*$ (Figure 13.11a),

$$\sigma_{av.} = (1 - x^*/L)\sigma_\infty. \tag{13.15}$$

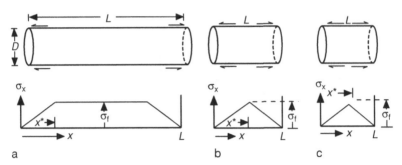

Figure 13.11. The distribution of fiber stress for three cases: $L > L^*$, $L = L^*$, and $L < L^*$.

If $L = L^*$ (Figure 13.11b),

$$\sigma_{av.} = (1/2)\sigma_{\infty}. \tag{13.16}$$

If $L < L^*$ (Figure 13.11c), σ_x never reaches σ_{∞}, and the average value of axial stress is

$$\sigma_{av.} = L/(2L^*)\sigma_{\infty}. \tag{13.17}$$

The composite modulus and strength can be calculated by substituting σ_{av} for σ_f in Equation 13.1.

The problem with this analysis is that the shear stress, τ, between matrix and fiber is not constant from $x = 0$ to $x = x^*$ along the length of the fiber. The shear stress, τ, between the fiber and the matrix decreases with x as the stress (and therefore the elastic elongation) of the fiber increases. It is reasonable to assume that τ increases as the difference between the strains in the fiber and matrix increases. In turn, this difference is proportional to $\sigma_{\infty} - \sigma_x$ because the fiber strain at x is σ_x/E_f and the matrix strain is σ_{∞}/E_f. If it is assumed that $\tau = C(\sigma_f - \sigma_x)$, where C is a constant, in Equation 13.10 and integrating,

$$\sigma_x = \sigma_{\infty}[1 - \exp(-2Cx/D)]. \tag{13.18}$$

This leads to the stress distribution sketched in Figure 13.12 near the end of the fiber. For long fibers ($L > 2x^*$), the average load carried by the fiber is not much different from that calculated with the simpler analysis.

A still more rigorous analysis results in an expression for the shear stress,

$$\tau = E_f(D/4)\exp(\beta)\sinh[\beta(L/2 - x)]/\cosh[\beta(L/2)], \tag{13.19}$$

where β is given by

$$\beta = (1/4)(G_m/E_f)/\ln(\phi/V_f). \tag{13.20}$$

Here ϕ is the ideal packing factor for the arrangements of fibers. For a square array, $\phi = \pi/4 = 0.785$, and for a hexagonal array $\phi = \pi/(2\sqrt{3}) = 0.907$. As

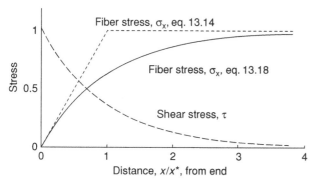

Figure 13.12. A more realistic stress distribution near the end of a fiber.

the ratio of the matrix shear modulus to the fiber tensile modulus increases, the load is transferred from the matrix to the fibers over a shorter distance.

Failures with Discontinuous Fibers

Failure may occur either by fracture of fibers or by the fibers pulling out of the matrix. Both possibilities are shown in Figure 13.13. If as a crack in the matrix approaches the fiber the plane of the crack is near the end of the fiber, pullout will occur. If it is not near the end, the fiber will fracture. Figure 13.14 is a picture showing the pullout of boron fibers in an aluminum matrix. To fracture a fiber of diameter, D, the force, F, must be

$$F = \sigma^* \pi D^2/4, \tag{13.21}$$

where σ^* is the fracture strength of the fiber.

Figure 13.13. Sketch showing some fibers fracturing at the crack and others pulling out.

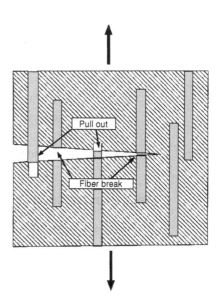

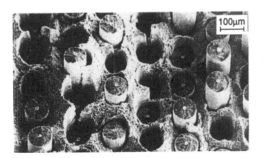

Figure 13.14. Photograph of SiC fibers pulling out of a titanium matrix. From T. W. Clyne and P. J. Withers, *An introduction to Metal Matrix Composites*, Cambridge U. Press (1993).

Much more energy is absorbed if the fibers pull out. The fiber pull-out force is

$$F = \tau^* \pi D x, \tag{13.22}$$

where τ^* is the shear strength of the fiber–matrix interface. Fibers will pull out if $\tau^* \pi D x$ is less than the force to break the fibers, $\sigma^* \pi D^2/4$. The critical pullout distance, x^*, corresponds to the two forces being equal,

$$x^* = (\sigma^*/\tau^*)D/4. \tag{13.23}$$

Fibers of length L less than $2x^*$ will pull out, and the average pull-out distance will be $L/4$. Therefore, the energy, U_{po}, to pull out the fiber is the integral of $\tau^* \pi D \int x \mathrm{d}x$ between limits of 0 and $L/4$,

$$U_{po} = \tau^* \pi D L^2/32. \tag{13.24}$$

However, if the fiber length is greater than $2x^*$, the probability that it will pull out is $2x^*/L$ and the average pull-out distance is $x^*/2$, so the average energy expended in pull-out is

$$U_{po} = \tau^* \pi D(x^*/2)^2(2x^*/L) = t^* \pi D x^{*3}/(2L). \tag{13.25}$$

Comparison of Equations 13.24 and 13.25 show that the pull-out energy, and hence the toughness, increase with L up to a critical length, $L = 2x^* = (\sigma^*/\tau^*)D/2$. Further increase of L decreases the fracture toughness, as shown in Figure 13.15. Because the composite stiffness and strength continue to increase with increasing fiber length, it is often desirable to decrease the fiber-matrix shear strength, τ^*, and thereby increase x^* so longer fibers can be used without decreasing toughness.

Failure Under Compression

In compression parallel to aligned fibers, failures can occur by fiber buckling. This involves lateral shearing of the matrix between fibers, and therefore the compressive strength of the composite depends on the shear moduli. Yielding of composites with polymeric fibers is sensitive to buckling under compression

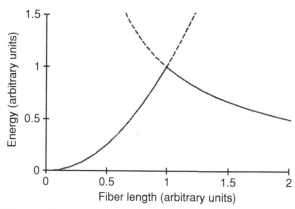

Figure 13.15. The energy expended in fiber pullout increases with fiber length up to a critical length and then decreases with further length increase.

because of the low compressive strengths of the fibers themselves. Figure 13.16 shows in-phase buckling of a Kevlar fiber-reinforced composite in compression. The low strength of some polymer fibers in compression was discussed in Chapter 12.

Typical Properties

Two types of polymer matrixes are common: epoxy and polyester. Most of the polymers used for matrix materials have moduli of 2 to 3 GPa and tensile strengths in the range of 35 to 70 MPa. Fiber reinforcements include glass, boron, Kevlar, and carbon. Properties of some epoxy matrix composite systems are given in Table 13.2 and some of the commonly used fibers in Table 13.3.

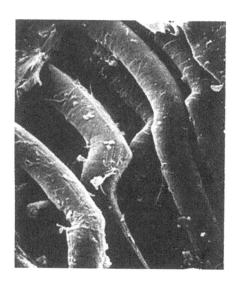

Figure 13.16. Buckling of Kevlar fibers in an epoxy matrix under compression. From *Engineered Materials Handbook, v.*1, *Composites*, ASM International (1987).

Table 13.2. *Properties of epoxy matrix composites*

Fiber	Vol % fiber	Young's modulus (GPa)		Tensile strength (MPa)	
		Longitudinal	Transverse	Longitudinal	Transverse
E-glass unidirectional	60	40	10	780	28
E-glass bidirectional	35	16.5	16.5	280	280
E-glass chopped matte	20	7	7	100	100
Boron unidirectional	60	215	24	1400	65
Kevlar29 unidirectional	60	50	5	1350	–
Kevlar49 unidirectional	60	76	6	1350	30
Carbon	62	145		1850	

Other fiber composites include ceramics reinforced with metal or ceramic fibers. Metals such as aluminum-base alloys may be reinforced with ceramic fibers to increase their stiffness. In some eutectic systems, directional solidification can lead to rods of one phase reinforcing the matrix.

Particulate Composites

Composites reinforced by particles rather than long fibers include such diverse materials as concrete (cement-paste matrix with sand and gravel particles), polymers filled with wood flour, and "carbide tools" with a cobalt-base matrix alloy hardened by tungsten carbide particles. Sometimes the purpose is simply economics, for example, wood flour is cheaper than plastics. Another objective

Table 13.3. *Typical fiber properties*

Fiber	Young's modulus (GPa)	Tensile strength (GPa)	Elongation (%)
carbon (PAN HS)	250	2.7	1
carbon (PAN HM)	390	2.2	0.5
SiC	70		
steel	210	2.5	
E-glass	70	1.75	
B	390	2 to 6	
Kevlar 29	65	2.8	4
Kevlar 49	125	2.8	2.3
Al_2O_3	379	1.4	
β-SiC	430	3.5	

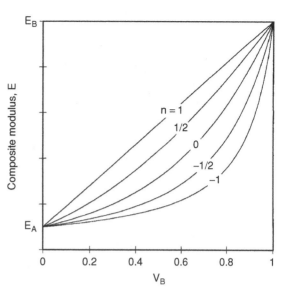

Figure 13.17. Dependence of Young's modulus on volume fraction according to Equation 13.26 for several values of n. The subscripts A and B represent the lower and higher modulus phases. The isostrain and isostress models correspond to $n = +1$ and -1. Here the ratio of $E_B/E_A = 10$.

may be increased hardness, such as carbide tools. The isostrain and isostress models (Equations 13.2 and 13.3) are upper and lower bounds for the dependence of Young's modulus on volume fraction. The behavior of particulate composites is intermediate and can be represented by a generalized rule of mixtures of the form

$$E^n = V_A E_A^n + V_B E_B^n, \tag{13.26}$$

where A and B refer to the two phases. The exponent, n, lies between the extremes of $n = +1$ for the isostrain model and $n = -1$ for the isostress model. Figure 13.17 shows the dependence of E on volume fraction for several values of n. If the modulus of the continuous phase is much greater than that of the particles, n $= 0.5$ is a reasonable approximation. For high-modulus particles in a low-modulus matrix, n < 0 is a better approximation.

A simple model for a particulate composite with a large fraction of the harder phase is illustrated in Figure 13.18. The harder phase, A, is a series of cubes. Let the distance between cube centers be 1 and the thickness of the softer phase, B, be t. Then the volume fraction of A is $V_A = (1 - t)^3$. The composite can be considered as a series of columns of alternating A and B loaded in parallel with columns of B.

If the behavior of the columns of alternating A and B is described by the lower-bound Equation 13.3, the modulus, E', of these columns is

$$1/E' = (1 - t)/E_A + t/E_B \tag{13.27}$$

so a lower bound to the modulus of the overall composite can be found using Equation 13.2 as

$$E_{av} = (1 - t)^2 E' + [1 - (1 - t)^2] E_B. \tag{13.28}$$

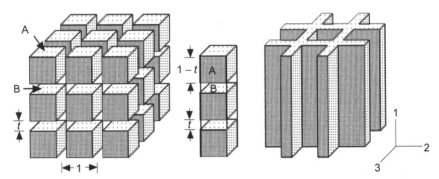

Figure 13.18. Brick-wall model of a composite. Hard particles in the shape of cubes are separated by a soft matrix of thickness, t.

An upper bound for the composite modulus, E_{ave}, can be found by treating these columns and the material between them with the isostrain model,

$$E_{av} = t(2 - t)E_B + (1 - t)^2 E'. \tag{13.29}$$

This is for this geometric arrangement and very much better solution than Equation 13.28. Figure 13.19 shows the results of calculations for $E_A = 3E_B$ and $\upsilon_A = \upsilon_B = 0.3$ using Equations 13.28 and 13.29. Also shown for comparison are the upper-bound and lower-bound equations.

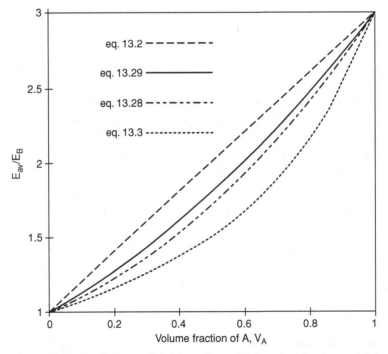

Figure 13.19. Predictions of brick-wall model for the elastic modulus of a particulate composite with $E_A = 3E_B$ and $\upsilon_A = \upsilon_B = 0.3$. Note that the predictions of Equations 13.28 and 13.29 are not very different.

EXAMPLE PROBLEM #13.2: Estimate the modulus for a composite consisting of 60% WC particles in a matrix of cobalt using the brick-wall model. The Young's moduli for WC and Co are 700 GPa and 210 GPa, respectively. WC is the discontinuous phase, A.

Solution: $t = 1 - 0.6^{1/3} = 0.156$. From Equation 13.29, $\sigma_{2B}/\sigma_1 = 0.615$

From Equation 13.30, $\varepsilon_{1B}/\sigma_1 = 2.81 \times 10^{-3}\,\mathrm{GPa}^{-1}$,
From Equation 13.31, $\varepsilon_{1A}/\sigma_1 = 1.32 \times 10^{-3}\,\mathrm{GPa}^{-1}$,
From Equation 13.32, $E' = 644\,\mathrm{GPa}$,
From Equation 13.28, $E_{av} = 519\,\mathrm{GPa}$.

Lamellar Composites

Two or more sheets of materials bonded together can be considered as lamellar composites. Examples include safety glass, plywood, plated metals, and glazed ceramics. Consider sheets of two materials, A and B, bonded together in the x-y plane and loaded in this plane. The basic equations governing the strain are

$$\varepsilon_{xA} = \varepsilon_{xB}$$
$$\varepsilon_{yA} = \varepsilon_{yB}. \tag{13.30}$$

If both materials are isotropic and the loading is elastic, Equations 13.30 become

$$\varepsilon_{xA} = (1/E_A)(\sigma_{xA} - \upsilon_A\sigma_{yA}) = \varepsilon_{xB} = (1/E_B)(\sigma_{xB} - \upsilon_B\sigma_{yB})$$
$$\varepsilon_{yA} = (1/E_A)(\sigma_{yA} - \upsilon_A\sigma_{xA}) = \varepsilon_{yB} = (1/E_B)(\sigma_{yB} - \upsilon_B\sigma_{xB}). \tag{13.31}$$

The stresses are

$$\sigma_{xA}t_A + \sigma_{xB}t_B = \sigma_{xav},$$
$$\sigma_{yA}t_A + \sigma_{yB}t_B = \sigma_{yav},$$
$$\sigma_{zA} = \sigma_z = 0, \tag{13.32}$$

where t_A and t_B are the fractional thicknesses of A and B.

Now consider loading under uniaxial tension applied in the x-direction. Substituting $\sigma_{yav} = 0, \sigma_{yB} = -(t_A/t_B)\sigma_{yA}$ and $\sigma_{xB} = (\sigma_{xav} - t_A\sigma_{xA})/t_B$ into Equations 13.31,

$$\sigma_{xA} - \upsilon_A\sigma_{yA} = (E_A/E_B)[(\sigma_{xav} - t_A\sigma_{xA})/t_B + \upsilon_B(t_A/t_B)\sigma_{yA}] \quad \text{and}$$
$$\sigma_{yA} - \upsilon_A\sigma_{xA} = (E_A/E_B)[(V_A/V_B)\sigma_{yA} + \upsilon_B(\sigma_{xav} - t_A\sigma_{xA})/t_B]. \tag{13.33}$$

Young's modulus according to an upper-bound isostrain model for loading in the plane of the sheet can be expressed as $E = N/D$, where

$$N = E_B^2 t_B^2 (1 - \upsilon_A{}^2) + 2 E_A E_B t_A t_B (1 - \upsilon_A \upsilon_B) + E_A{}^2 t_A^2 (1 - \upsilon_B^2) \quad \text{and}$$
$$D = E_B t_B (1 - \upsilon_A^2) + E_A t_A (1 - \upsilon_B{}^2). \tag{13.34}$$

For the special case in which $\upsilon_B = \upsilon_A$, these expressions reduce to the upper-bound model,

$$E = E_B t_B + E_A t_A. \tag{13.35}$$

EXAMPLE PROBLEM #13.3: Calculate the composite modulus for a sandwich of two sheets of fiber-reinforced polyester (each 0.5 mm thick) surrounding rubber (4 mm thick). For the fiber-reinforced polyester, $E_B = 7$ GPa and $\upsilon_B = 0.3$. For the rubber, $E_A = 0.25$ GPa and $\upsilon_A = 0.5$.

Solution: Substituting $t_A = 0.8$ and $t_B = 0.2$ into Equation 13.35

$N = (7 \times 0.2)^2 (0.75) + 2(7)(0.25)(0.2)(0.8)(0.85) + (0.25)(0.8)^2 (0.91) = 2.21$
$D = (7 \times 0.2)(0.75) + (0.25 \times 0.8)(0.91) = 1.232$
$E = N/D = 2.21/1.232 = 1.79$ GPa

Foams

There are two types of foams: closed cell foams and open cell (or reticulated) foams. In open foams, air or other fluids are free to circulate. These are used for filters and as skeletons. They are often made by collapsing the walls of closed cell foams. Closed cell foams are much stiffer and stronger than open cell foams because compression is partially resisted by increased air pressure inside the cells. Figure 13.20 shows that the geometry of open and closed cell foams can modeled by Kelvin tetrakaidecahedra.

The elastic stiffness depends on the relative density. In general, the dependence of relative stiffness, E^*/E, where E^* is the elastic modulus of the structure and E is the modulus of the solid material on relative density, is of the form

$$E^*/E = (\rho^*/\rho_s)^n. \tag{13.36}$$

Experimental results shown in Figure 13.21 indicate that for open cells $n = 2$, so

$$E^*/E = (\rho^*/\rho_s)^2. \tag{13.37}$$

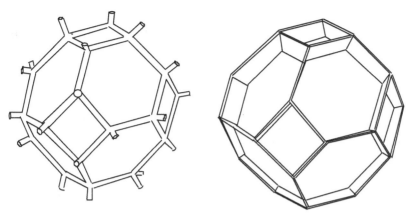

Figure 13.20. Open and closed cell foams modeled by tetrakaidecahedra. From W. F. Hosford, *Materials for Engineers*, Cambridge U. Press (2008).

For closed cell foams, E^*/E is much greater and $n < 2$. Although deformation under compression of open cell foams is primarily by ligament bending, compression of closed cell wall foams involves gas compression and wall stretching in addition to wall bending.

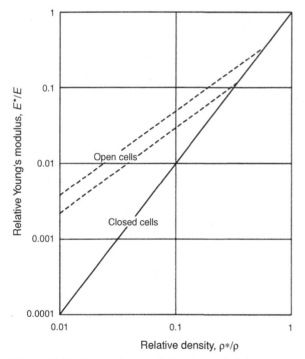

Figure 13.21. Dependence of elastic moduli on the density of open and closed cell foams.

Metal foams are useful in many engineering applications because of their extremely light weight, good energy absorption, high ratios of strength and stiffness to weight, and outstanding damping capability.

Notes

The engineering use of composites dates back to antiquity. The Bible, in Exodus, cites the use of straw in clay bricks (presumably to prevent cracking during fast drying under the hot sun in Egypt. The Mayans and Incas incorporated vegetable fibers into their pottery. English builders used sticks in the "wattle and daub" construction of timber frame buildings. In colonial days, hair was used to strengthen plaster, the hair content being limited to 2% or 3% to insure workability.

Saracens made composite bows composed of a central core of wood with animal tendons glued on the tension side and horn on the compression side. These animal products are able to store more elastic energy in tension and compression respectively than wood.

Shortly after Leo Baekelund (1863–1944) developed phenolformaldehyde ("Bakelite") in 1906, he found that adding fibers to the resin greatly increased its toughness. The first use of this molding compound was for the gearshift knob of the 1916 Rolls Royce.

Widespread commercial use of glass-reinforced polymers began soon after World War II. Freshly formed glass fibers of 0.005 mm to 0.008 mm diameter have strengths up to 4 GPa, but the strength is greatly reduced if the fibers contact one another. For that reason, they are coated with an organic compound before being bundled together.

REFERENCES

K. K. Chawla, Composite Materials, Science and Engineering, Springer-Verlag (1978).

P. K. Mallick, Fiber-reinforced *Composites, Materials, Manufacturing & Design*, Dekker (1988).

Engineered Materials Handbook, Vol I, Composites, ASM International (1987).

N. C. Hilyard, *Mechanics of Cellular Plastics*, MacMillan (1982).

E. Andrew, W. Sanders and L. J. Gibson, Compressive and tensile behaviour of aluminium foams, *Mater Sci Eng* A 270 (1999).

L. J. Gibson & M. F. Ashby *Cellular Foams*, Cambridge (1999).

A. Kim, M. A. Hasan, S. H. Nahm and S. S. Cho, *Composite Structures*, v. 71 (2005).

Problems

1. What would be the critical length, L^*, for maximum load in a 10-μm diameter fiber with a fracture strength of 2 GPa embedded in a matrix such that the shear strength of the matrix-fiber interface is 100 MPa?

Table 13.4. *Properties of steel wire and aluminum*

	Young's modulus	Yield strength (GPa)	Poisson's ratio (MPa)	Linear coef. of thermal expans (K^{-1})
aluminum	70	65	0.3	24×10^{-6}
steel	210	280	0.3	12×10^{-6}

2. Estimate the greatest value of the elastic modulus that can be obtained by long randomly oriented fibers of E-glass embedded in an epoxy resin if the volume fraction is 40%. Assume the modulus of the epoxy is 5 GPa.

3. Carbide cutting tools are composites of very hard tungsten carbide particles in a cobalt matrix. The elastic moduli of tungsten carbide and cobalt are 102×106 and 30×106 psi, respectively. It was experimentally found that the elastic modulus of a composite containing 52 volume percent carbide was 60×106 psi. What value of the exponent, n, in Equation 13.26 would this measurement suggest? A trial-and-error solution is necessary to solve this. (Note that n = 0 is a trivial solution.)

4. A steel wire (1.0 mm diameter) is coated with aluminum, 0.20 mm thick. (See Table 13.4)

   A. Will the steel or the aluminum yield first as tension is applied to the wire?

   B. What tensile load can the wire withstand without yielding

   C. What is the composite elastic modulus?

   D. Calculate the composite thermal expansion coefficient.

5. Consider a carbon-reinforced epoxy composite containing 45 volume percent unidirectionally aligned carbon fibers (see Table 13.5). A. Calculate the composite modulus. B. Calculate the composite tensile strength. Assume both the epoxy and carbon are elastic to fracture.

Table 13.5. *Properties of epoxy and carbon fibers*

	Young's modulus	Tensile strength
epoxy	3 GPa	55 MPa
carbon	250 GPa	2.5 GPa

14 Mechanical Working

Introduction

The shapes of most metallic products are achieved by mechanical working. The exceptions are those produced by casting and by powder processing. Mechanical shaping processes are conveniently divided into two groups, bulk forming and sheet forming. Bulk forming processes include rolling, extrusion, rod and wire drawing, and forging. In these processes, the stresses that deform the material are largely compressive. One engineering concern is to ensure that the forming forces are not excessive. Another is ensuring that the deformation is as uniform as possible so as to minimize internal and residual stresses. Forming limits of the material are set by the ductility of the work piece and by the imposed stress state.

Products as diverse as cartridge cases, beverage cans, automobile bodies, and canoe hulls are formed from flat sheets by drawing or stamping. In sheet forming, the stresses are usually tensile and the forming limits usually correspond to local necking of the material. If the stresses become compressive, buckling or wrinkling will limit the process.

Bulk Forming Energy Balance

An energy balance is a simple way of estimating the forces required in many bulk-forming processes. As a rod or wire is drawn through a die, the total work, W_t, equals the drawing force, F_d, multiplied by the length of wire drawn, ΔL, $W_t = F_d \Delta L$. Expressing the drawing force as $F_d = \sigma_d A$, where A is the area of the drawn wire and σ_d is the stress on the drawn wire, $W_t = \sigma_d A \Delta L$ (Figure 14.1). As $A \Delta L$ is the drawn volume, the actual work per volume, w_t, is

$$w_t = \sigma_d. \tag{14.1}$$

The total work per volume can also be expressed as the sum of the individual work terms,

$$w_t = w_i + w_f + w_r. \tag{14.2}$$

Figure 14.1. Drawing of a wire. The force, F_d working through a distance of ΔL deforms a volume $A\Delta L$.

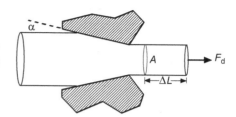

The *ideal work,* w_i, is the work that would be required by an ideal process to create the same shape change as the real process. The ideal process is an imaginary process in which there is no friction and no redundancy. It does not matter whether such a process is possible or not. For example, the ideal work to draw a wire from 10 mm to 7 mm diameter can be found by considering uniform stretching in tension, even though in reality the wire would neck. In this case, the ideal work/volume is

$$w_i = \int \sigma \, d\varepsilon. \qquad (14.3)$$

For a material that does not work harden, this reduces to

$$w_i = \sigma \varepsilon. \qquad (14.4)$$

The work of friction between the wire-drawing die and the wire can also be expressed on a per-volume basis, w_f. In general, the total frictional work, w_f, is roughly proportional to the flow stress, σ, of the wire and to the area of contact, A_c, between the wire and the die. This is because the normal force between the wire and die is $F_n = \sigma_n A_c$, where the contact pressure between tool and die, σ_n, depends on the flow stress, σ. The frictional force is $F_f = \mu F_n$, where μ is the coefficient of friction. For a constant reduction, the contact area, A_c, increases as the die angle, α, decreases so w_f increases as the die angle, α, decreases. A simple approximation for low die angles is

$$w_f = w_i(1 + \varepsilon/2)\mu\cot\alpha. \qquad (14.5)$$

This predicts that the frictional work increases with increasing reduction and decreasing die angle.

The redundant work is the energy expended in plastic straining that is not required by the ideal process. During drawing, streamlines are bent as they enter the die and again as they leave the die. This involves plastic deformation that is not required in the ideal process of pure stretching. It also causes shearing of the surface relative to the interior, as schematically illustrated in Figure 14.2. The redundant work per volume, w_r, increases with die angle, α. A simple approximation for relatively low die angles is

$$w_r = (2/3)\sigma\tan\alpha. \qquad (14.6)$$

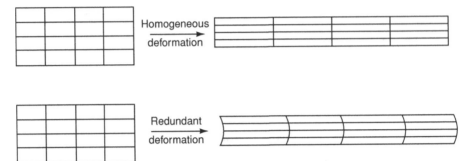

Figure 14.2. Redundant deformation involves shearing of surface relative to the interior.

This equation predicts that w_r increases with increasing die angle but does not depend on the total strain, ε. Therefore, the ratio w_r/w_i decreases with increasing reduction.

A mechanical efficiency, η, can be defined such that

$$\eta = w_i/w_t. \tag{14.7}$$

Note that η is always less than 1. For wire drawing, typical drawing efficiencies are in the range of 50% to 65%.

Figure 14.3 shows how each of the work terms and the efficiency depend on die angle for a fixed reduction. There is an optimum die angle, α^*, for which the efficiency is a maximum. Figure 14.4 shows that in general, with increased reduction the efficiency increases and that the optimum die angle also increases. The reduction is defined as $r = (A_o - A)/A_o$.

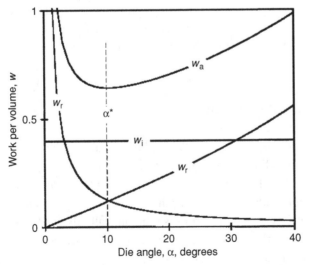

Figure 14.3. Dependence of w_a, w_i, w_r, and w_f on die angle, α. Note that there is an optimum die angle, α^*, for which the total work per volume is a minimum.

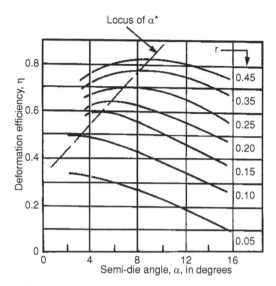

Figure 14.4. Variation of efficiency with die angle and reduction. Note that the efficiency and the optimum die angle, α^*, increase with reduction. From W. F. Hosford and R. M. Caddell, *Metal Forming: Mechanics and Metallurgy 3rd Ed.*, Cambridge U. Press (2007). Adapted from data of J. Wistreich, *Metals Rev.*, v. 3 (1958).

For wire drawing, there is a maximum reduction that can be made in a single drawing pass. If a greater reduction is attempted, the stress on the drawn section will exceed its tensile strength and the wire will break instead of drawing through the die. Therefore, multiple passes are required to make wire. After the first few dies, little additional work-hardening occurs so the flow stress, σ, becomes the tensile strength. The limit then can be expressed as $\sigma = \sigma_d = w_t$. Substituting Equation 14.7 ($w_a = w_i/h$) and Equation 14.4 ($w_i = \sigma\varepsilon$), the drawing limit corresponds to

$$\varepsilon = \eta. \tag{14.8}$$

The diameter reduction, $(D_o - D_f)/D_o = 1 - D_f/D_o$, can be expressed as $\Delta D/D_o = 1 - \exp(-\varepsilon/2)$. The maximum diameter reduction for an efficiency of 50% is then $\Delta D/D_o = 1 - \exp(0.5/2) = 22\%$, and for an efficiency of 65% it is 28%. Because the reductions per pass are low in wire drawing, the optimum die angles are also very low, as suggested by Figure 14.4.

A work balance for extrusion is very similar, except now $W_t = F_{ext}\Delta L_o$ where F_{ext} is the extrusion force and A_o and ΔL_o are the cross-sectional area and length of billet extruded (Figure 14.5). The actual work per volume is $w_t = W_t/(A_o\Delta L_o) = F_{ext}/A_o$. This can be expressed as

$$w_t = P_{ext}, \tag{14.9}$$

where P_{ext} is the extrusion pressure. Because material is being pushed through the die instead of being pulled, there is no inherent limit to the possible reduction per pass as there is in drawing. Therefore, extrusions are made in a single operation with extrusion ratios, A_o/A_f, as large as 16 or more. Because of the large reductions, large die angles are common. Often dies with 90° die angles

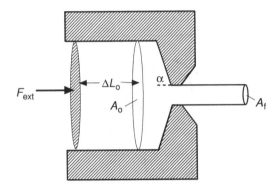

Figure 14.5. Direct extrusion. The extrusion force, F_{ext}, working through a distance, ΔL, deforms a volume, $A_o \Delta L$.

are used because they permit more of the billet to be extruded before the die is opened. Similar work balances can be made for rolling and forging.

EXAMPLE PROBLEM #14.1: Consider the 16:1 extrusion of a round aluminum billet from a diameter of 24 cm to a diameter of 6 cm. Assume that the flow stress is 15 MPa at the appropriate temperature and extrusion rate. Estimate the extrusion force, assuming an efficiency of 0.50. Calculate the lateral pressure on the side of the extrusion chamber.

Solution: $P_{ext} = (1/\eta)\sigma\varepsilon$. Substituting $\eta = 0.50$, $\sigma = 15$ MPa and $\varepsilon = \ln(16) = 2.77$, $P_{ext} = (1/.5)(15)(2.77) = 83.2$ MPa. $F_{ext} = (\pi D^2/4)P_{ext} = 3.76$ MN.

For axially symmetric flow, the radial stress on the billet, σ_r, and the hoop stress, σ_c, must be equal and they must be large enough so the material in the chamber is at its yield stress. Using von Mises, $(\sigma_2 - \sigma_3)^2 + (\sigma_3 - \sigma_1)^2 + (\sigma_1 - \sigma_2)^2 = 2Y^2$. With $\sigma_2 = \sigma_3 = \sigma_r$, $\sigma_r - \sigma_1 = -Y$, $\sigma_r = -Y + \sigma_1$. Substituting $\sigma_1 = -P_{ext} = -83.2$ MN, and $Y = 15$ MPa, $\sigma_r = -98.2$ MPa.

EXAMPLE PROBLEM #14.2: Find the horsepower that would be required to cold roll a 48-in. wide sheet from a thickness of 0.030 in. to 0.025 in. if the exit speed is 80 ft/s and the flow stress is 10,000 psi. Assume a deformation efficiency of 80% and neglect work-hardening.

Solution: The rate of doing work is (work/volume)(volume rolled/time) = $[(1/\eta)\sigma\varepsilon](vwt) = (1/0.80)(10 \times 10^3 \text{lb/in.}^2)$ $[\ln(0.030/0.025)](80 \times 12\text{in./s})$ $(48\text{ in.})(0.025\text{ in.}) = 5.8 \times 10^6 \text{in.-lbs/s} = (5.8 \times 10^6/12 \text{ ft.lbs/s})[1.8 \times 10^{-3}\text{hp/} (\text{ft.lbs/s})] = 876$ hp. This is the horsepower that must be delivered by the rolls to the metal. (This solution neglects energy losses between the motor and the rolls.)

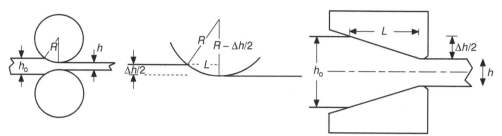

Figure 14.6. Deformation zones in rolling (left) and extrusion and drawing (right). The parameter, Δ, is defined as the ratio of the mean height or thickness to the contact length.

Deformation Zone Geometry

One of the chief concerns with mechanical working is the homogeneity of the deformation. Inhomogeneity affects the hardness distribution, residual stress patterns, and internal porosity in the final product and the tendency to crack during forming. In turn, the inhomogeneity is dependent on the geometry of the deformation zone, which can be characterized in various processes by a parameter, Δ, defined as

$$\Delta = h/L. \tag{14.10}$$

Here h is the height or thickness or diameter at the middle of the deformation zone and L is the length of contact between tools and work piece. Several examples are shown in Figure 14.6. For rolling, $R^2 = L^2 + (R - \Delta h/2)^2 = L^2 + R^2 - 2R\Delta h + \Delta h^2/4$ and $L \approx \sqrt{(R^2 \Delta h)}$, so

$$\Delta \approx h/\sqrt{(R^2 \Delta h)}, \tag{14.11}$$

and for extrusion and drawing,

$$\Delta = h/L = h \tan \alpha/(\Delta h/2). \tag{14.12}$$

EXAMPLE PROBLEM #14.3: In rolling of a sheet of 2.0 mm thickness with 15 cm diameter rolls, how large a reduction (percent reduction of thickness) would be necessary to insure $\Delta \geq 1$?

Solution: Take $\Delta = h/(R\Delta h)^{1/2} = h/(Rrh_o)^{1/2}$ and assume for the purpose of calculation that $h = h_o$. Then $r = (h_o/R)/\Delta^2 = (2.0/75)/\Delta^2$, so for $\Delta = 1, r = 0.027$ or 2.7%. (The reduction is small enough that the assumption that $h = h_o$ is justified.)

One way of describing the inhomogeneity is by a *redundant work factor*, Φ, defined as

$$\Phi = 1 + w_r/w_i. \tag{14.13}$$

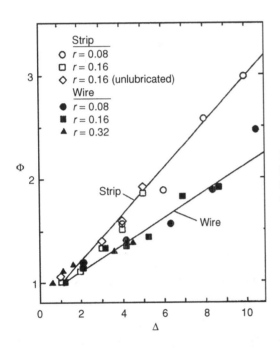

Figure 14.7. The increase of the redundancy factor, Φ, with Δ for both strip and wire drawing. Note that friction has little effect on Φ. From Hosford and Caddell, *ibid.*

Figure 14.7 shows the increase of Φ with Δ for both strip drawing (plane-strain) and wire drawing (axially symmetric flow). Friction has little effect on redundant strain.

Another way of characterizing inhomogeneity is by an *inhomogeneity factor*, defined as

$$IF = H_s / H_c. \tag{14.14}$$

where H_s and H_c refer to the hardnesses at the surface and at the center. The inhomogeneity factor increases with increased die angle, α as shown in Figure 14.8. With large die angles, shearing at the surface causes more work-hardening, therefore increasing *IF*.

As Δ increases, forming processes leave the surface under increasing residual tension, as shown in Figure 14.9. Under high Δ conditions, a state of hydrostatic tension develops near the centerline, and this may cause pores to open around inclusions (Figure 14.10). If the conditions are extreme, these pores may grow to form macroscopic cavities at the centerline, as shown in Figure 14.11.

Friction in Bulk Forming

In forging, friction can play a very important role when the length of contact, L, between tools and workpiece is large compared with the workpiece thickness, h. During compression of a slab between two parallel platens, the

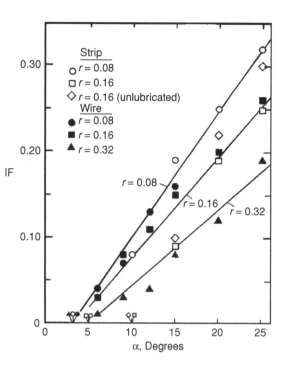

Figure 14.8. The increase of the inhomo-geneity factor, $IF = 1 + H_{surf}/H_{center}$ with die angle. From Hosford and Caddell. Data from J. J. Burke, ScD Thesis, MIT (1968).

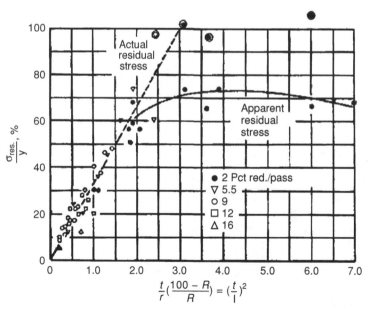

Figure 14.9. Residual stresses at the surface of cold-rolled brass strips increase with $\Delta = t/l$. From W. M. Baldwin, *Proc. ASTM*, v. 49 (1949).

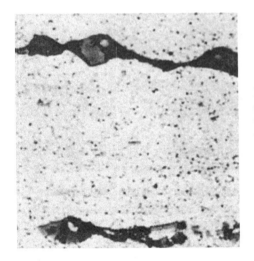

Figure 14.10. Voids opening at inclusions near the centerline during drawing under high-D conditions. With high-Δ, the center is under hydrostatic tension. From H. C. Rogers, R. C. Leach and L. F. Coffin Jr., *Final Report Contract Now-65–0097-6*, Bur. Naval Weapons (1965).

friction tends to suppress lateral flow of the work material. Higher compressive stresses are necessary to overcome this restraint. The greater the L/h is, the greater the average compressive stress needs to be.

For compression in plane-strain with a constant coefficient of friction, the ratio of the average pressure to the plane-strain flow strength P_{av}/σ_o, is

$$P_{av}/\sigma_o = (h/\mu L)[\exp(\mu L/h) - 1], \tag{14.15}$$

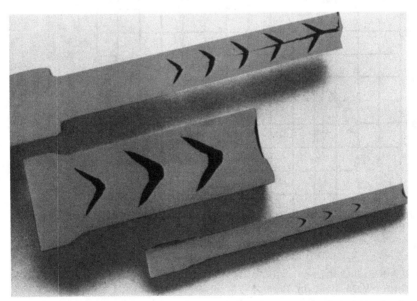

Figure 14.11. Centerline cracks in extruded steel bars. Note that the reductions were very small, so Δ was very large. From D. J. Blickwede, *Metals Progress*, v. 97, (May 1970).

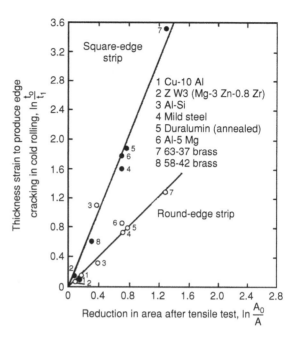

Figure 14.12. Correlation of the strain at which edge cracking occurs in flat rolling with the reduction of area in a tension test. The greater fracture strains of the "square-edge" strips reflects a greater state of compression at the edge during rolling. From M. G. Cockcroft and D. J. Latham, *J. Inst. Metals*, v. 96 (1968).

which can be approximated by $P_{av}/\sigma_o \approx 1 + \mu L/h$ for small values of $\mu L/h$. For compression of circular discs of diameter, D, the ratio of the average pressure to the flow strength P_{av}/Y, is

$$P_{av}/Y = 2(h/\mu D)^2[\exp(\mu D/h) - \mu D/h - 1]. \qquad (14.16)$$

For small values of $\mu L/D$, this can be approximated by

$$P_{av}/Y = 1 + (1/3)\mu D\Delta/h + (1/12)(\mu D\Delta/h)^2. \qquad (14.17)$$

In the rolling of thin sheets, the pressure between the rolls and workpiece can become so large that the rolls bend. Such bending would produce a sheet that is thicker in the middle than at the edges. To prevent this, the rolls may be backed up with larger rolls or ground to a barrel-shape to compensate for bending.

Formability

In bulk forming processes other than wire drawing, formability is limited by fracture. Whether fracture occurs or not depends on both the material and the process. The formability of a material is related to its reduction of area in a tension test. A material with large fracture strain is likely to have a large formability. Figure 14.12 shows the correlation of edge cracking during rolling with tensile ductility for a number of materials. The factors governing a material's fracture strain were discussed in Chapter 9. Large inclusion content and high strength tend to decrease ductility and formability.

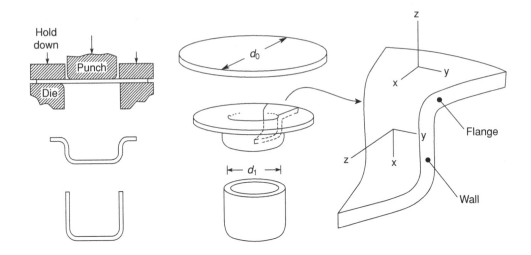

Figure 14.13. Schematic illustration of cup drawing showing the coordinate axes. As the punch descends, the outer circumference must undergo compression so that it will be small enough to flow over the die lip. From Hosford and Caddell, *ibid*.

Formability also depends on the level of hydrostatic stress in the process. The difference in Figure 14.12 between the fracture strains in "square edge" and "round edge" strips reflect this. If the edge of a rolled strip is allowed to become round during rolling, the through-thickness compressive stress is less at the edge. Therefore, a greater tensile stress in the rolling direction is needed to cause the same elongation as in the middle of the strip. With the greater level of hydrostatic tension, the strain to fracture is lower.

Deep Drawing

A major concern in sheet forming is tensile failure by necking. Sheet forming operations may be divided into *drawing*, where one of the principal strains in the plane of the sheet is compressive, and *stretching* where both of the principal strains in the plane of the sheet are tensile. Compressive stresses normal to the sheet are usually negligible in both cases.

A typical drawing process is the making of cylindrical, flat-bottom cups. It starts with a circular disc blanked from a sheet. The blank is placed over a die with a circular hole and a punch forces the blank to flow into the die cavity, as sketched in Figure 14.13. A hold-down force is necessary to keep the flange from wrinkling. As the punch descends, the blank is deformed into a hat shape and finally into a cup. Deforming the flange consumes most of the energy. The energy expended in friction and some in bending and unbending as material flows over the die lip is much less. The stresses in the flange are compressive in the hoop direction and tensile in the radial direction. The tension is a maximum at the inner lip and the compression a maximum at the outer periphery.

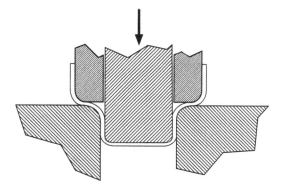

Figure 14.14. Direct redrawing. A sleeve around the punch acts as a hold-down while the punch descends.

As with wire drawing, there is a limit to the amount of reduction that can be achieved. If the ratio of the initial blank diameter, d_o, to the punch diameter, d_1, is too large, the tensile stress required to draw the material into the die will exceed the tensile strength of the wall, and the wall will fail by necking. It can be shown that for an isotropic material, the largest ratio of d_o/d_1 that can be drawn (*limiting drawing ratio* or LDR) is

$$(d_o/d_1)_{max} = \exp(\eta), \qquad (14.18)$$

where η is the deformation efficiency. For an efficiency of 70%, $(d_o/d_1)_{max} = 2.01$. There is very little thickening or thinning of the sheet during drawing, so this corresponds to cup with a height-to-diameter ratio of 3/4. Materials with R-values greater than unity have somewhat greater limiting drawing ratios. This is because with a large R-value, the increased thinning resistance permits larger wall stresses before necking and as well as easier flow in the plane of the sheet, which decreases the forces required. Forming cylindrical cups with a greater height-to-diameter ratio requires *redrawing*, as shown in Figure 14.14. In can making, an additional operation called *ironing* thins and elongates the walls, as illustrated in Figure 14.15.

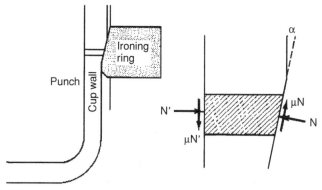

Figure 14.15. Section of a cup wall and ironing ring during ironing. The die ring causes wall thinning. Friction on opposite sides of the wall acts in opposing directions. From Hosford and Caddell, *ibid.*

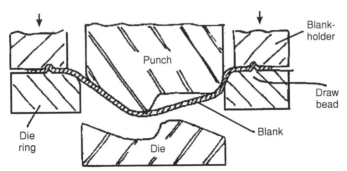

Figure 14.16. Sketch of a sheet stamping operation by Duncan. The sheet is stretched to conform to the tools rather than being squeezed between them. In this case, the lower die contacts the sheet and causes a reverse bending only after it has been stretched by the upper die. For many parts, there is not a bottom die.

Stamping

Operations variously called *stamping*, *pressing*, or even *drawing* involve clamping the edges of the sheet and forcing it into a die cavity by a punch, as shown in Figure 14.16. The metal is not squeezed between tools. Rather, it is made to conform to the shape of the tools by stretching. Failures occur by either *wrinkling* or by *localized necking*. Wrinkling will occur if the restraint at the edges is not great enough to prevent excessive material being drawn into the die cavity. Blank-holder pressure and draw beads are often used to control the flow of material into the die. If there is too much restraint, more stretching may be required to form the part than the material can withstand. The result is that there is a window of permissible restraint for any part, as illustrated in Figure 14.17. With too little blank-holder force, the depth of draw

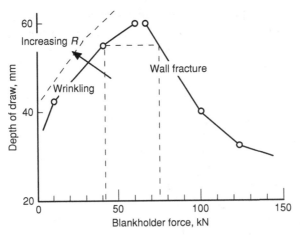

Figure 14.17. The effect of blankholder force on the possible depth of draw. If the blankholder force is too low, wrinkling will result from too much material being drawn into the die cavity. Too great a blankholder force will require too much stretching of the sheet and result in a necking failure. For deep draws, there may be only a narrow window of permissible blankholder forces. Large R-values widen this window.

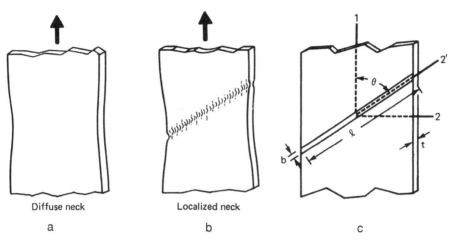

Figure 14.18. Development of (a) a diffuse neck and (b) a localized neck. The coordinate axes used in the analysis are shown in (c). The characteristic angle, θ, of the local neck must be such that $\varepsilon_{2'} = 0$. From Hosford and Caddell, *ibid.*

is limited by wrinkling. On the other hand, if the blank-holder force is too large, too little material is drawn into the die cavity to form the part and the part fails by *localized necking*. The size of this window depends on properties of the sheet. Materials with greater strain-hardening exponents, n, can stretch more before necking failure so the right-hand limit is raised and shifted to the left. There will be more lateral contraction in materials having a large R-value, thus decreasing the wrinkling tendency. Therefore, large R-values increase the wrinkling limit and shifts it to the left.

In a tension test of a ductile material, the maximum load and *diffuse necking* occur when $\sigma = d\sigma/d\varepsilon$. For a material that follows a power-law ($\sigma = K\varepsilon^n$) stress-strain curve, this is when $\varepsilon = n$. This diffuse necking occurs by local contraction in both the width and thickness directions and is generally not a limitation in practical sheet forming. If the specimen is wide (as a sheet is), such localization must be very gradual. Eventually, a point is reached where lateral contraction in the plane of the sheet ceases. At this point, a *localized neck* forms in which there is only thinning. In uniaxial tension, the conditions for localized necking are $\sigma = 2d\sigma/d\varepsilon$, or $\varepsilon = 2n$. Figure 14.18 illustrates general and localized necking in a tensile specimen. The characteristic angle at which the neck forms must be such that the incremental strain in that direction, $d\varepsilon_\theta$, becomes zero.

Because $d\varepsilon_{2'} = d\varepsilon_1\cos^2\theta + d\varepsilon_2\sin^2\theta$ and $d\varepsilon_{2'} = 0$,

$$\tan\theta = (-d\varepsilon_1/d\varepsilon_2)^{1/2}. \qquad (14.19)$$

For uniaxial tension of an isotropic material, $d\varepsilon_2 = -d\varepsilon_1/2$ so $\tan\theta = \sqrt{2}$ and $\theta = 54.7°$.

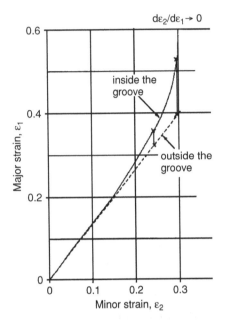

Figure 14.19. Calculated strain paths inside and outside of a pre-existing groove for a material with $n = 0.22$, $m = 0.012$, and $R = 1.5$. The initial thickness in the groove was assumed to be 0.5% less than outside. A strain path of $\varepsilon_2 = 0.75\varepsilon_1$ was imposed outside the groove. As $d\varepsilon_2/d\varepsilon_1 \to 0$ inside the groove, a local neck develops and deformation outside the groove virtually ceases, fixing a limit strain of $\varepsilon_1 = 0.4$ and $\varepsilon_2 = 0.3$.

EXAMPLE PROBLEM #14.4: Find the angle between the tensile axis and the local neck form in a tension test on material with a strain ratio $R = 2$.

Solution: $R = \varepsilon_3/\varepsilon_2 = -(\varepsilon_1 + \varepsilon_2)/\varepsilon_2 = -\varepsilon_1/\varepsilon_2 - 1$ so $-\varepsilon_1/\varepsilon_2 = 1 + R$. Substituting into Equation 14.19, $\tan\theta = (1 + R)^{1/2} = \sqrt{3}$ $\theta = 60°$.

In sheet forming, ε_2 will be less negative if there is a tensile stress, σ_2. This in turn will increase the characteristic angle. For plane-strain conditions, $d\varepsilon_2 = 0$, $\theta = 90°$.

If $d\varepsilon_2$ is positive, there is no angle at which a local neck can form. However, under conditions of biaxial stretching a small pre-existing groove perpendicular to the largest principal stress can grow gradually into a localized neck. The strain ε_2 must be the same inside and outside the groove. However, the stress σ_1 within the groove will be greater than outside the groove, so the strain ε_1 will be also be larger in the groove. As the strain rate $\dot\varepsilon_1$ inside the groove accelerates, the ratio of $\dot\varepsilon_2/\dot\varepsilon_1$ within the groove approaches zero, which is the condition necessary for local necking. Figure 14.19 shows how the strain path inside and outside a groove can diverge. Straining outside the groove will virtually cease once $\dot\varepsilon_2/\dot\varepsilon_1$ becomes very large. The terminal strain outside the groove is the *limit strain*. Very shallow grooves are sufficient to cause such localization. How rapidly this happens depends largely on the strain-hardening exponent, n, and to a lesser extent the strain-rate exponent, m.

A plot of the combinations of strains that lead to necking failure is called a *forming limit diagram* or (FLD). Figure 14.20 is such a plot for low carbon steels. Combinations of strains below the forming limits are safe, whereas

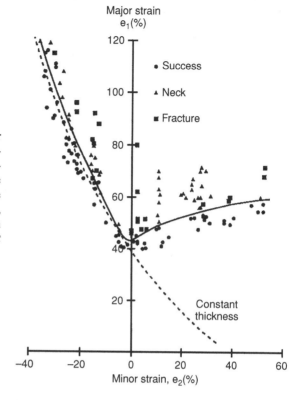

Figure 14.20. Forming limit diagram for low-carbon steel. The strain combinations below the curve are acceptable, whereas those above it will cause local necking. The limiting strains here are expressed as engineering strains, although true strains could have been plotted. Data from S. S. Hecker, *Sheet Metal Ind.*, v. 52 (1975).

those above the limits will cause local necking. Note that the lowest failure strains correspond to plane strain, $\varepsilon_2 = 0$.

Some materials, when stretched in biaxial tension, may fail by shear fracture instead of local necking. Shear failures are also possible before necking under large strains in the left-hand side of the diagram. It should be noted that if the minor strain, ε_2, is less than $-\varepsilon_1/2$, the minor stress, σ_2, must be compressive. Under these conditions, wrinkling or buckling of the sheet may occur. Because wrinkled parts are usually rejected, this too should be regarded as a failure mode. The possibilities of both shear fracture and wrinkling are shown in Figure 14.21.

The strains vary from one place to another in a given part. A pan being formed is sketched in Figure 14.22. The strain paths at several different locations are indicated schematically on a forming limit diagram.

Spinning (Figure 14.23) is a sheet forming process that is suitable for forming axially symmetric parts. A tool forces a disc that is spinning parallel to it axis of rotation to conform to a mandrel. For pure shear, the deformation is restricted to just under the tool, so no deformation occurs elsewhere. This eliminates danger wrinkling. Because tooling costs are low and the process is relatively slow, spinning is suited to producing low production parts.

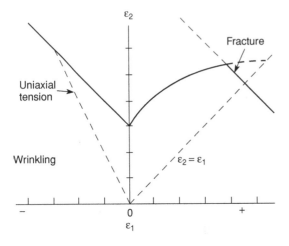

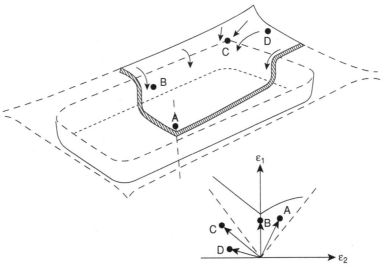

Figure 14.21. Schematic forming limit diagram showing regions where wrinkling may occur and a possible fracture limit in biaxial tension.

Figure 14.22. Sketch showing the strain paths in several locations during the drawing of a pan. At A, the strain state is nearly balanced biaxial tension. The deformation at point B is in plane strain. At C, there is drawing with contraction in the minor strain direction. At D, the there may be enough compression in the 2-direction to cause wrinkling. Courtesy of J. L. Duncan.

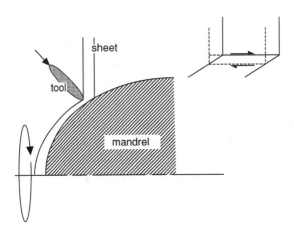

Figure 14.23. Sketch of a spinning operation. If the tool causes only shearing parallel to the axis of rotation, no deformation occurs in the flange.

Notes

The first aluminum two-piece beverage cans were produced in 1963. They replaced the earlier steel cans made from three separate pieces: a bottom, the wall (which was bent into a cylinder and welded), and a top. The typical beverage can is made from a circular blank, 5.5 in. in diameter, by drawing it into a 3.5-in. diameter cup, redrawing to 2.625 in. diameter, and then ironing the walls to achieve the desired height. There are about 200 billion made in the United States each year. Beverage cans account for about one-fifth of the total usage of aluminum.

In flat rolling of thin sheets or foils, the L/h ratio can become very large. This raises the average roll pressure to the extent that rolls elastically flatten where they are in contact with the work material. This flattening prevents further thinning. One way of overcoming this difficulty is to roll two foils at the same time, effectively doubling h. One side of commercial aluminum foil has a matte finish, whereas the other side is shiny. The sides with the matte finish were in contact during rolling. Another method of circumventing the roll-flattening problem is to use a cluster mill (Figure 14.24), developed by Sendzimer. The work roll has a very small diameter to keep L/h low. To prevent bending of the work rolls, both are backed up by two rolls of somewhat

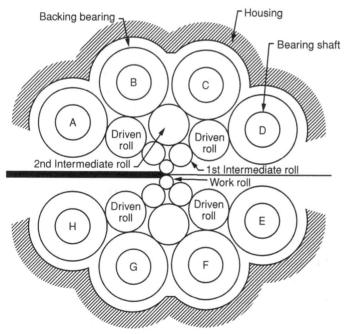

Figure 14.24. A Sendzimer mill. Small-diameter work rolls are used keep the ratio of L/h low. A cluster of larger backup rolls prevent bending of the work rolls. Courtesy of T. Sendizimer.

larger diameter. These are, in turn, backed up by three still larger diameter rolls, and so on.

REFERENCES

W. F. Hosford and R. M. Caddell, *Metal Forming: Mechanics and Metallurgy, 3rd Ed.* Cambridge U. Press, (2007).
W. A. Backofen, *Deformation Processing*, Addison Wesley (1972).
E. M. Mielnick, *Metalworking Science and Engineering*, McGraw-Hill (1991).
Z. Marciniak and J. L. Duncan, *Mechanics of Sheet Forming*, Edward Arnold (1992).
G. E. Dieter, *Mechanical Metallurgy, 2nd Ed.* McGraw-Hill (1976).

Problems

1. A small special alloy shop received an order for slabs 4 inches wide and 1/2 inch thick of an experimental superalloy. The shop cast ingots 4 in. × 4 in. × 12 in. and hot rolled them in a 12 in. diameter mill, making reductions of about 5% per pass. On the fifth pass, the first slab split longitudinally parallel to the rolling plane. The project engineer, the shop foreman, and a consultant met to discuss the problem. The consultant proposed applying forward and back tension during rolling, the project engineer suggested reducing the reduction per pass, and the shop foreman favors greater reductions per pass. With whom would you agree? Explain your reasoning.

2. A high-strength steel bar must be cold reduced from a diameter of 1.00 in. to 0.65 in. A number of schedules have been proposed. Which of the schedules would you choose to avoid drawing failure and minimize the likelihood of centerline bursts? Explain. Assume $\eta = 0.50$.

    A. A single reduction in a die having a die angle of 8°.

    B. Two passes (1.00 to 0.81 in. and 0.81 to 0.65 in.) using dies with angles of $\alpha = 8°$.

    C. Three passes (1.00 to 0.87 in. and 0.87 to 0.75 in, and 0.75 to 0.65 in.) using dies with angles of 8°.

    D., E., and F. Same schedules as A, B, and C, except using dies with $\alpha = 15°$.

3. Your company is planning to produce niobium wire and you have been asked to decide how many passes would be required to reduce the wire from 0.125 to 0.010 inches in diameter. In laboratory experiments with dies having the same angle as will be used in the operation, it was found that the efficiency increased with reduction, $\eta = 0.65 + \Delta\varepsilon/3$, where $\Delta\varepsilon$ is the strain in the pass. Assume that in practice the efficiency will be only 75% of that found in the laboratory experiments. To insure no failures, stress on the drawn section of wire must never exceed 80% of its strength. Neglect work hardening.

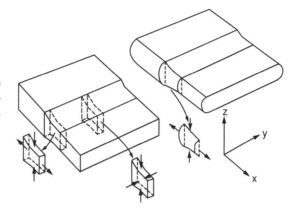

Figure 14.25. The difference between the stress states at the edges of square-edge and round-edge strips during rolling. From Hosford and Caddell, *ibid.*

4. One stand of a hot-rolling mill is being designed. It will reduce 60-in. wide sheet from 0.150 to 0.120 in. thickness at an exit speed of 20 feet per second. Assume that the flow stress of the steel at the temperature and strain rate in the rolling mill is 1500 psi. If the deformation efficiency is 82% and the efficiency of transferring energy from the motor to the mill is 85%, what horsepower motor should be used?

5. A typical aluminum beverage can is 2.6 in. in diameter and 4.8 in. tall. The thickness of the bottom is 0.010 in. and the wall thickness is 0.004. The cans are produced from circular blanks 0.010 in. thick by drawing, redrawing, and ironing to a height of 5.25 in. before trimming.
   A. Calculate the diameter of the initial circular blank.
   B. Calculate the total effective strain at the top of the cup from rolling, drawing, redrawing and ironing.

6. Figure 14.12 shows that large reductions in rolling can be achieved before edge cracking occurs if the edges are maintained square instead of being allowed become rounded. Figure 14.25 shows the edge elements. Explain in terms of the stress state at the edge why the greater strains are possible with square edges.

7. When aluminum alloy 6061-T6 is cold drawn through a series of dies with a 25% reduction per pass, a loss of density is noted, as shown in Figure 14.26. Explain why the density loss increases with larger angle dies.

8. Assuming that in the drawing of cups, the thickness of the cup bottom and wall is the same as that of the original sheet, and find an expression for the ratio of the cup height to diameter, h/d_1, in terms of the ratio of blank diameter to cup diameter, d_o/d_1. Evaluate h/d_1 for $d_o/d_1 = 1.5$, 1.75, 2.0, and 2.25 and plot h/d_1 vs. d_o/d_1.

9. In the drawing of cups with a conical wall, the elements between the punch and the die must deform in such a way that their circumference shrinks, otherwise they will buckle or wrinkle. The tendency to wrinkle can be decreased by applying a greater blankholder force, as shown in

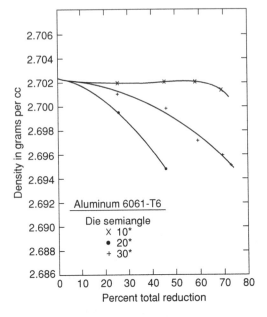

Figure 14.26. Density changes in aluminum alloy 6161-T6 during drawing. From H. C. Rogers, *General Electric Co. Report No. 69-C-260* (1969).

Figure 14.27. This increases the radial tension between the punch and die. How would the R-value of the material affect how much blankholder force is necessary to prevent wrinkling?

10. Figure 14.28 is a forming limit diagram for a low-carbon steel. This curve represents the combinations of strains that would lead to failure under plane stress ($\sigma_3 = 0$) loading.

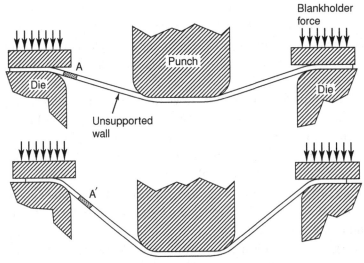

Figure 14.27. Drawing of a conical cup. As element A is drawn into the die cavity, its circumference must shrink. This requires enough tensile stretching in the radial direction. From Hosford and Caddell, *ibid.*

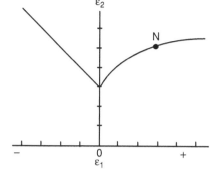

Figure 14.28. Forming limit diagram for a low-carbon steel.

A. Show the straining path inside a Marciniak defect under biaxial tension that would lead to necking at point N.
B. Plot carefully on the diagram the strain path that corresponds to uniaxial tension ($\sigma_3 = 0$).
C. Describe how this path would be changed for a material with a value of $R > 1$.

11. Consider drawing a copper wire from 0.125 to 0.100 in. diameter. Assume that $\sigma = (55\text{MPa})\varepsilon^{0.36}$ in a die for which $\alpha = 6°$.

A. Calculate the drawing strain.
B. Calculate the reduction of area.
C. Use Figure 14.4 to determine η and calculate the drawing stress.

Calculate the yield strength of the drawn wire. Can this reduction be made?

15 Anisotropy

Introduction

Although it is frequently assumed that materials are isotropic (i.e., have the same properties in all directions) they rarely are. In single crystals elastic properties vary with crystallographic direction. There are two principal causes of anisotropic mechanical behavior of crystalline materials. The anisotropy of elastic and plastic properties are caused by crystallographic texture (preferred crystallographic orientation of grains.) Anisotropic plastic behavior is principally a result of alignment of grain boundaries or second-phase inclusions by prior mechanical working. Alignment of molecules in polymers causes elastic, plastic, and fracture behavior.

Elastic Anisotropy

Hooke's law for anistropic materials can be expressed in terms of compliances, s_{ijmn}, which relate the contributions of individual stress components to individual strain components. In the most general case, Hooke's law may be written as

$$e_{ij} = s_{ijmn}\sigma_{mn}, \tag{15.1}$$

where summation is implied. The individual compliances, s_{ijmn}, form a fourth-order tensor. However, this relation is greatly simplified by taking advantage of the relations $\sigma_{ij} = \sigma_{ji}$ and $\gamma_{ij} = 2e_{ij} = 2e_{ji}$ and by adopting a new subscript convention for the compliances: Two subscripts, i and j, are used to identify the location (row and column of the compliance in the matrix) rather than the axes of the stress and strain components. Because the subscripts identify the location of the compliances, the order of the stress and strain components is critical. With this new convention, Hooke's law becomes

$$e_{11} = s_{11}\sigma_{11} + s_{12}\sigma_{22} + s_{13}\sigma_{33} + s_{14}\sigma_{23} + s_{15}\sigma_{31} + s_{16}\sigma_{12}$$
$$e_{22} = s_{21}\sigma_{11} + s_{22}\sigma_{22} + s_{23}\sigma_{33} + s_{24}\sigma_{23} + s_{25}\sigma_{31} + s_{26}\sigma_{12}$$
$$e_{33} = s_{31}\sigma_{11} + s_{32}\sigma_{22} + s_{33}\sigma_{33} + s_{34}\sigma_{23} + s_{35}\sigma_{31} + s_{36}\sigma_{12}$$
$$\gamma_{23} = s_{41}\sigma_{11} + s_{42}\sigma_{22} + s_{43}\sigma_{33} + s_{44}\sigma_{23} + s_{45}\sigma_{31} + s_{46}\sigma_{12} \tag{15.2}$$
$$\gamma_{31} = s_{51}\sigma_{11} + s_{52}\sigma_{22} + s_{53}\sigma_{33} + s_{54}\sigma_{23} + s_{55}\sigma_{31} + s_{56}\sigma_{12}$$
$$\gamma_{12} = s_{61}\sigma_{11} + s_{62}\sigma_{22} + s_{63}\sigma_{33} + s_{64}\sigma_{23} + s_{65}\sigma_{31} + s_{66}\sigma_{12}.$$

Note that the subscripts 1, 2, and 3 on the stress and strain terms refer to crystallographic axes. It can be shown for all combinations of i and j that $s_{ij} = s_{ji}$, This simplifies the matrix to

$$e_{11} = s_{11}\sigma_{11} + s_{12}\sigma_{22} + s_{13}\sigma_{33} + s_{14}\sigma_{23} + s_{15}\sigma_{31} + s_{16}\sigma_{12}$$
$$e_{22} = s_{12}\sigma_{11} + s_{22}\sigma_{22} + s_{23}\sigma_{33} + s_{24}\sigma_{23} + s_{25}\sigma_{31} + s_{26}\sigma_{12}$$
$$e_{33} = s_{13}\sigma_{11} + s_{23}\sigma_{22} + s_{33}\sigma_{33} + s_{34}\sigma_{23} + s_{35}\sigma_{31} + s_{36}\sigma_{12}$$
$$\gamma_{23} = s_{14}\sigma_{11} + s_{24}\sigma_{22} + s_{34}\sigma_{33} + s_{44}\sigma_{23} + s_{45}\sigma_{31} + s_{46}\sigma_{12} \tag{15.3}$$
$$\gamma_{31} = s_{15}\sigma_{11} + s_{25}\sigma_{22} + s_{35}\sigma_{33} + s_{45}\sigma_{23} + s_{55}\sigma_{31} + s_{56}\sigma_{12}$$
$$\gamma_{12} = s_{16}\sigma_{11} + s_{26}\sigma_{22} + s_{36}\sigma_{33} + s_{46}\sigma_{23} + s_{56}\sigma_{31} + s_{66}\sigma_{12}.$$

Thus, in the most general case there are 21 independent elastic constants. Symmetry about the x, y, and z axes of most materials causes many of the constants to be equal or disappear. With orthotropic symmetry, the 23, 31, and 12 planes are planes of mirror symmetry so that the 1, 2, and 3 axes are axes of twofold rotational symmetry. Paper, wood, and rolled sheets of metal have such symmetry. In this case, Hooke's law simplifies to

$$e_{11} = s_{11}\sigma_{11} + s_{12}\sigma_{22} + s_{13}\sigma_{33}$$
$$e_{22} = s_{12}\sigma_{11} + s_{22}\sigma_{22} + s_{23}\sigma_{33}$$
$$e_{33} = s_{13}\sigma_{11} + s_{23}\sigma_{22} + s_{33}\sigma_{33}$$
$$\gamma_{23} = s_{44}\sigma_{23} \tag{15.4}$$
$$\gamma_{31} = s_{55}\sigma_{31}$$
$$\gamma_{12} = s_{66}\sigma_{12}.$$

For cubic crystals, the 1, 2, and 3 axes are equivalent <100> directions so $s_{11} = s_{12} = s_{13}$, $s_{12} = s_{13} = s_{14}$, and so $s_{44} = s_{55} = s_{66}$ and the matrix further simplifies to

$$e_{11} = s_{11}\sigma_{11} + s_{12}\sigma_{22} + s_{12}\sigma_{33}$$
$$e_{22} = s_{12}\sigma_{11} + s_{11}\sigma_{22} + s_{12}\sigma_{33}$$
$$e_{33} = s_{12}\sigma_{11} + s_{12}\sigma_{22} + s_{11}\sigma_{33}$$
$$\gamma_{23} = s_{44}\sigma_{23} \tag{15.5}$$
$$\gamma_{31} = s_{44}\sigma_{31}$$
$$\gamma_{12} = s_{44}\sigma_{12}.$$

Table 15.1. *Elastic compliances* $(TPa)^{-1}$ *for various cubic crystals**

Material	s_{11}	s_{12}	s_{44}
Cr	3.10	−0.46	10.10
Fe	7.56	−2.78	8.59
Mo	2.90	−0.816	8.21
Nb	6.50	−2.23	35.44
Ta	6.89	−2.57	12.11
W	2.45	−0.69	6.22
Ag	22.26	−9.48	22.03
Al	15.82	−5.73	35.34
Cu	15.25	−6.39	8.05
Ni	7.75	−2.98	8.05
Pb	94.57	−43.56	67.11
Pd	13.63	−5.95	13.94
MgO	4.05	−0.94	6.60
MnO	7.19	−2.52	12.66
NaCl	22.80	−4.66	78.62

* From W. Boas and J. M. MacKenzie, *Progress in Metal Physics*, v.2, 1950

For hexagonal and cylindrical symmetry, z is the hexagonal (or cylindrical) axis. The 1 and 2 axes are any pair of axes perpendicular to zz. In this case $s_{13} = s_{23}, s_{22} = s_{33}$, and $s_{66} = 2(s_{11} - s_{12})$, which simplifies the matrix to

$$e_{11} = s_{11}\sigma_{11} + s_{12}\sigma_{22} + s_{13}\sigma_{33}$$
$$e_{22} = s_{12}\sigma_{11} + s_{11}\sigma_{22} + s_{13}\sigma_{33}$$
$$e_{33} = s_{13}\sigma_{11} + s_{13}\sigma_{22} + s_{33}\sigma_{33}$$
$$\gamma_{23} = s_{44}\sigma_{23}$$
$$\gamma_{31} = s_{44}\sigma_{31}$$
$$\gamma_{12} = 2(s_{11} - s_{12})\sigma_{12}. \tag{15.6}$$

The elastic compliances for several cubic crystals are listed in Table 15.1.

> **EXAMPLE PROBLEM #15.1:** Derive expressions for Young's modulus and Poison's ratio for a cubic crystal stresses in uniaxial tension along the [112] direction.
>
> *Solution:* Let $x = [112]$. The y and z axes must be normal to each other and to x. Let them be $y = [11\bar{1}]$ and $z = [1\bar{1}0]$. The direction cosines between these axes and the cubic axes of the crystal can be found by dot products; for example, $\ell_{1x} = [100] \cdot [112] = (1 \cdot 1 + 0 \cdot 1 + 0 \cdot 2)/[(1^2 + 0^2 + 0^2)(1^2 + 1^2 + 2^2)] = 1/\sqrt{6}$. Forming Table 15.2 of direction cosines,

Table 15.2. *Direction cosines*

	$x = [112]$	$y = [11\bar{1}]$	$z = [1\bar{1}0]$
$1 = [100]$	$1/\sqrt{6}$	$1/\sqrt{3}$	$1/\sqrt{2}$
$2 = [010]$	$1/\sqrt{6}$	$1/\sqrt{3}$	$-1/\sqrt{2}$
$3 = [001]$	$2/\sqrt{6}$	$-1/\sqrt{3}$	0

For uniaxial tension along [112], $\sigma_y = \sigma_z = \sigma_{yz} = \sigma_{zx} = \sigma_{yxy} = 0$, so

$$\sigma_1 = (1/6)\sigma_x, \sigma_2 = (1/6)\sigma_x, \sigma_3 = (2/3)\sigma_x, \sigma_{23} = (1/3)\sigma_x, \sigma_{31} = (1/3)\sigma_x,$$
$$\sigma_{12} = (1/6)\sigma_x.$$

Using Equations 15.5,

$$e_1/\sigma_x = (1/6)s_{11} + (1/6)s_{12} + (2/3)s_{12} + (1/6)s_{11} = (1/6)s_{11} + (5/6)s_{12},$$
$$e_2/\sigma_x = (1/6)s_{11} + (1/6)s_{12} + (2/3)s_{12} + (1/6)s_{11} = (1/6)s_{11} + (5/6)s_{12},$$
$$e_3/\sigma_x = (1/6)s_{11} + (1/6)s_{12} + (2/3)s_{12} + (1/6)s_{11} = (2/3)s_{11} + (1/3)s_{12},$$
$$\gamma_{23}/\sigma_x = \gamma_{31}/\sigma_x = (1/3)s_{44}, \gamma_{12}/\sigma_x = (1/6)s_{44},$$

Transforming these strains back onto the x, y, and z axes,

$$e_x = (1/6)e_1 + (1/6)e_2 + (2/3)e_3 + (1/3)\gamma_{23} + (1/3)\gamma_{31} + (1/6)\gamma_{12}$$
$$= [(1/6)s_{11} + (1/2)s_{12} + (1/4)s_{44}]/\sigma_x$$
$$e_y = (1/3)e_1 + (1/3)e_2 + (1/3)e_3 + (1/3)\gamma_{23} + (1/3)\gamma_{31} + (1/3)\gamma_{12}$$
$$= [(1/3)s_{11} + (2/3)s_{12} + (1/6)s_{44}]/\sigma_x$$
$$e_z = (1/2)e_1 + (1/2)e_2 - (1/2)\gamma_{23}$$
$$= [(1/6)s_{11} + (5/6)s_{12} + (1/6)s_{44}]/\sigma_x.$$

Young's modulus is $E_{112} = (e_x/\sigma_x)^{-1} = [(1/6)s_{11} + (1/2)s_{12} + (1/4)s_{44}]^{-1}$.

There are two Poison's ratios, $\upsilon_y = -e_y/e_x$ and $\upsilon_z = -e_z/e_x$

$$\upsilon_y = -e_y/e_x = -[(1/3)s_{11} + (2/3)s_{12} + (1/6)s_{44}]/[(1/6)s_{11} + (1/2)s_{12} + (1/4)s_{44}]$$
$$= -(s_{11} + 8s_{12} + 2s_{44})/(2s_{11} + 6s_{12} + 3s_{44})$$

$$\upsilon_z = -e_z/e_x = -[(1/6)s_{11} + (5/6)s_{12} + (1/6)s_{44}]/[(1/6)s_{11} + (1/2)s_{12} + (1/4)s_{44}]$$
$$= -(2s_{11} + 10s_{12} - 2s_{44})/(2s_{11} + 6s_{12} + 3s_{44})$$

For fcc and bcc metals the extremes of E are in the [100] and [111] directions. Table 15.3 shows this variation for some metals.

The basic forms of the elastic constant matrix can be used for materials that are not single crystals but which have similar symmetries of structure and properties. Such materials may be either natural or artificial composites. Examples include:

1. Orthotropic (same as orthorhombic) symmetry (three axes of two-fold symmetry):

Table 15.3. *Values of $E_{[111]}/E_{[100]}$ for several cubic crystals*

Material	$E_{[111]}/E_{[100]}$	Material	$E_{[111]}/E_{[100]}$
Cr	0.76	Fe	2.14
Mo	0.92	Nb	0.52
W	1.01	Al	1.20
Cu	2.91	Ni	2.36
Si	1.44	MgO	1.39
KCl	0.44	NaCl	0.74
ZnS	2.17	GaAs	1.66

   a. Rolled sheets or plates, both before and after recrystallization. The anisotropy is caused by crystallographic texture. The axes are the rolling, transverse, and thickness directions.
   b. Tubes and swaged wire or rod having cylindrical crystallographic textures. The axes are the radial, tangential (hoop), and axial directions.
   c. Wood. The axes are the radial, tangential and axial directions.
2. Tetragonal symmetry (one axis of four-fold symmetry, two axes of two-fold symmetry):
   a. Plywood with equal number of plies at 0° and 90°. The four-fold axis is normal to the sheet and the two two-fold axes are parallel to the wood grain in the two plies. The anisotropy is from oriented fibers.
   b. Woven cloth, window screens and composites reinforced with woven material if the cloth or screen has the same weft and warp (the two directions of thread in a woven cloth). The four-fold axis is normal to the cloth and the two two-fold axes are parallel to the threads of the weft and warp
3. Axial symmetry (same as hexagonal symmetry):
   a. Drawn or extruded wire and rod. The anisotropy is caused by a fiber texture, all directions normal to the wire axis being equivalent.
   b. Electroplates and portions of ingots with columnar grains. All directions normal to the growth direction are equivalent.
   c. Uniaxially aligned composites. The axis of symmetry is parallel to the fibers.

In principle, it should be possible to obtain the elastic moduli for a polycrystal from a weighted average of the elastic behavior of all orientations of crystals present in the polycrystal. However, the appropriate way to average is not obvious.

Thermal Expansion

The coefficients of thermal expansion in bcc and fcc crystals do not depend on crystallographic direction. However, in materials that do not have a cubic

Table 15.4. *Coefficients of thermal expansion for some non-cubic crystals*

Material	Struct.	Temp. range (°C)	α_{11} (mm/m)/K	α_{22} (mm/m)/K	α_{33} (mm/m)/K
zirconium	hcp	300 to 900	5.7	$\alpha_{22} = \alpha_{11}$	11.4
zinc	hcp	0 to 100	15.0	$\alpha_{22} = \alpha_{11}$	61.5
magnesium	hcp	0 to 35	24.3	$\alpha_{22} = \alpha_{11}$	27.1
cadmium	hcp	at 0	19.1	$\alpha_{22} = \alpha_{11}$	54.3
titanium	hcp	0 to 700	11.0	$\alpha_{22} = \alpha_{11}$	12.8
tin	tetr	at 50	16.6	$\alpha_{22} = \alpha_{11}$	32.9
calcite	hex	0 to 85	−5.6	$\alpha_{22} = \alpha_{11}$	25.1
uranium	ortho	at 75	20.3	−1.4	22.2

crystal structure the coefficient of thermal expansion does depend on crystallographic direction. Table 15.4 lists the thermal expansion coefficients of some non-cubic crystals.

Anisotropic Plasticity

The angular variation of yield strength in many sheet materials is not large. However, such a lack of variation does not indicate that the material is isotropic. The parameter that is commonly used to characterize the anisotropy is the *strain ratio* or *R-value** (Figure 15.1). This is defined as the ratio, R, of the contractile strain in the width direction to that in the thickness direction during a tension test,

$$R = \varepsilon_w / \varepsilon_t. \tag{15.7}$$

If a material is isotropic, the width and thickness strains, ε_w and ε_t, are equal, so for an isotropic material $R = 1$. However, for most materials R is usually either greater or less than 1 in real sheet materials. Direct measurement of the thickness strain in thin sheets is inaccurate. Instead, ε_t is usually deduced from the width and length strains, ε_w and ε_l, assuming constancy of volume, $\varepsilon_t = -\varepsilon_w - \varepsilon_l$.

$$R = -\varepsilon_w / (\varepsilon_w + \varepsilon_l). \tag{15.8}$$

To avoid constraint from the shoulders, strains should be measured well away from the ends of the gauge section. Some workers suggest that the strains be measured when the total elongation is 15%, if this is less than the necking strain. The change of R during a tensile test is usually quite small, and the lateral strains at 15% elongation are great enough to be measured with accuracy.

* Some authors use the symbol, r, instead of R.

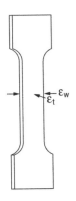 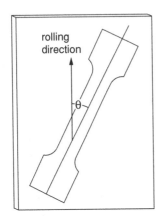

Figure 15.1. Tensile specimen cut from A sheet (left). The R-value is the ratio of the lateral strains, $\varepsilon_w/\varepsilon_t$, during the extension (right).

EXAMPLE PROBLEM #15.2: Consider the accuracy of R measurement for a material having an R of about 1.00. Assume the width is about 0.5 in and can be measured to an accuracy of ± 0.001 in., which corresponds to an uncertainty of strain of $\pm 0.001/0.5 = \pm 0.002$. The errors in measuring the length strain are much smaller. Estimate the error in finding R if the total strain were 5%. How would this be reduced if the total strain were 15%?

Solution: At an elongation of 5%, the lateral strain would be about 0.025, so the error in ε_w would be $\pm 0.002/0.025 = 8\%$. If the R-value were 1, this would cause an error in R of approximately $0.08 \times 0.5/(1 - 0.5)$, or 8%. At 15% elongation, the percent error should be about one third of this, or roughly $\pm 3\%$.

The value of R usually depends on the direction of testing. An average R-value is conventionally taken as

$$\bar{R} = (R_0 + R_{90} + 2R_{45})/4. \tag{15.9}$$

The angular variation of R is characterized by ΔR, defined as

$$\Delta R = (R_0 + R_{90} - 2R_{45})/2. \tag{15.10}$$

Both of these are important in analyzing what happens during sheet metal forming.

Although it is frequently assumed that materials are isotropic (have the same properties in all directions), they rarely are. There are two main causes of anisotropy. One cause is *preferred orientations* of grains or *crystallographic texture*. The second is *mechanical fibering*, which is the elongation and alignment of microstructural features such as inclusions and grain boundaries. Anisotropy of plastic behavior is almost entirely caused by the presence of preferred orientations.

The first complete quantitative treatment of plastic anisotropy was in 1948 by Hill,* who proposed an anisotropic yield criterion to accommodate such materials. It is a generalization of the von Mises criterion:

$$F(\sigma_y - \sigma_z)^2 + G(\sigma_z - \sigma_x)^2 + H(\sigma_x - \sigma_y)^2 + 2L\tau_{yz}^2 + 2M\tau_{zx}^2 + 2N\tau_{xy}^2 = 1, \tag{15.11}$$

where the axes x, y, and z are the symmetry axes of the material (e.g., the rolling, transverse, and through-thickness directions of a rolled sheet).

If the loading is such that the directions of principal stress coincide with the symmetry axes and if there is planar isotropy (properties do not vary with direction in the x-y plane), Equation 15.11 can be simplified to

$$(\sigma_y - \sigma_z)^2 + (\sigma_z - \sigma_x)^2 + R(\sigma_x - \sigma_y)^2 = (R+1)X^2, \tag{15.12}$$

where X is the yield strength in uniaxial tension.

For plane-stress ($\sigma_z = 0$), Equation 15.8 plots as an ellipse, as shown in Figure 15.2. The larger the value of R, the more the ellipse extends into the first quadrant. Thus, the strength under biaxial tension increases with R, as suggested earlier.

The corresponding flow rules are obtained by applying the general equation, $d\varepsilon_{ij} = d\lambda(\partial f/\partial\sigma_{ij})$ (Equation 5.14), where now $f(\sigma_{ij})$ is given by Equation 15.12:

$$d\varepsilon_x{:}\ d\varepsilon_y{:}\ d\varepsilon_z = [(R+1)\sigma_x - R\sigma_y - \sigma_z] : [(R+1)\sigma_y - R\sigma_x - \sigma_z] :$$
$$[2\sigma_z - \sigma_y - \sigma_x]. \tag{15.13}$$

This means that $\rho = \varepsilon_y/\varepsilon_x$ for $\sigma_z = 0$ and $\alpha = \sigma_y/\sigma_x$ for $\sigma_z = 0$, are related by

$$\rho = [(R+1)\alpha - R]/[(R+1) - R\alpha] \tag{15.14}$$

and

$$\alpha = [(R+1)\rho + R]/[(R+1) + R\rho]. \tag{15.15}$$

Note that if $R = 1$, these reduce to Equation 5.15, and the effective stress is

$$\bar{\sigma} = \{[(\sigma_y - \sigma_z)^2 + (\sigma_z - \sigma_x)^2 + R(\sigma_x - \sigma_y)^2]/(R+1)\}^{1/2}. \tag{15.16}$$

For $\sigma_z = 0$,

$$\bar{\sigma}/\sigma_x = \{[\alpha^2 + 1 + R(1 - \alpha)^2]/(R+1)\}^{1/2}. \tag{15.17}$$

The effective strain function is given by

$$\bar{\varepsilon}/\varepsilon_x = (\sigma_1/\bar{\sigma})(1 + \alpha\rho). \tag{15.18}$$

---

* R. Hill, *Proc Roy. Soc.* v.193A (1948).

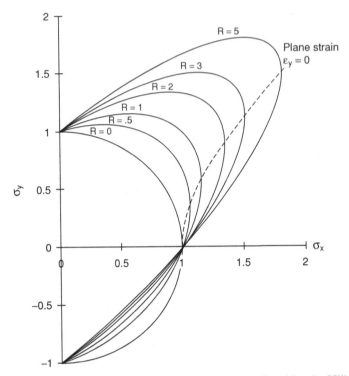

Figure 15.2. Plane stress ($\sigma_z = 0$) yield locus predicted by the Hill 1948 yield criterion for planar isotropy (Equation 5.34) with several values of R. The dashed line is the locus of stress states that produce plane strain ($\varepsilon_y = 0$). Note that the strength under biaxial tension increases with R. A large R indicates a resistance to thinning in a tension test, which is consistent with high strength under biaxial tension where thinning must occur. From W. F. Hosford, *The Mechanics of Crystals and Textured Polycrystals.* Used by permission of Oxford University Press (1993).

The Hill criterion often overestimates the effect of R-value on the flow stress. A modification of Equation 15.12, referred to as the high-exponent criterion, was suggested to overcome this difficulty,[*]

$$(\sigma_y - \sigma_z)^a + (\sigma_z - \sigma_x)^a + R(\sigma_x - \sigma_y)^a = (R+1)X^a, \qquad (15.19)$$

where a is an even exponent much larger than 2. Calculations based on crystallographic slip have suggested that a = 6 is appropriate for bcc metals and a = 8 for fcc metals. Figure 15.3 compares the yield loci predicted by this criterion and the Hill criterion for several levels of R.

With this criterion, the flow rules are

$$d\varepsilon_x : d\varepsilon_y : d\varepsilon_z = [R(\sigma_x - \sigma_y)^{a-1} + (\sigma_x - \sigma_z)^{a-1}] : [(\sigma_y - \sigma_z)^{a-1} + R(\sigma_y - \sigma_x)^{a-1}] :$$
$$[(\sigma_z - \sigma_x)^{a-1} + (\sigma_z - \sigma_y)^{a-1}]. \qquad (15.20)$$

---

[*] *W. F. Hosford, *7th North Amer. Metalworking Conf., SME*, Dearborn MI (1979) and R. W. Logan and W. F. Hosford, *Int. J. Mech. Sci.*, v.22, (1980).

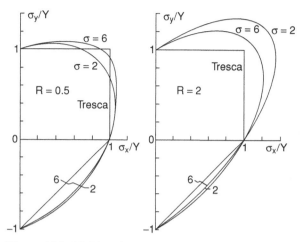

Figure 15.3. Yield loci predicted by the high-exponent criterion for planar isotropy. The loci for the 1948 Hill criterion correspond to $a = 2$. Note that with a large exponent, there is much less effect of R on strength under biaxial tension. From Hosford, *ibid.*

The effective stress function corresponding to Equation 15.19 is

$$\bar{\sigma} = [(\sigma_y - \sigma_z)^a + (\sigma_z - \sigma_x)^a + R(\sigma_x - \sigma_y)^a]/(R+1)\}^{1/a}. \tag{15.21}$$

The effective strain function is

$$\bar{\varepsilon} = (\sigma_x/\bar{\sigma})(1 + \alpha\rho), \tag{15.22}$$

where $\alpha = (\sigma_y - \sigma_z)/(\sigma_x - \sigma_y)$ and $\rho = \varepsilon_y/\varepsilon_x$.

Although the nonlinear flow rules for the high exponent criterion (Equation 15.20) cannot be explicitly solved for the stresses, iterative solutions with calculators or personal computers are simple. The nonquadratic yield criterion and accompanying flow rules (Equations 15.19 and 15.20 have been shown to give better fit to experimental data than the quadratic form (Equations 15.12 and 15.13).

EXAMPLE PROBLEM #15.3: A steel sheet, which has an R-value of 1.75 in all directions in the sheet, was stretched in biaxial tension with $\sigma_z = 0$. Strain measurements indicate that throughout the deformation, $\varepsilon_y = 0$. Find the stress ratio, $\alpha = \sigma_y/\sigma_x$, that prevailed according to the Hill yield criterion (Equation 15.8) and according to the non-quadratic yield criterion (Equation 15.15) with a = 8.

Solution: For the Hill criterion, Equation 15.8 gives $\alpha = [(R+1)\rho + R]/[R+1+R\rho]$. With $\rho = 0$, $\alpha = R/(R+1) = 1.75/2.75 = 0.636$.

For the non-quadratic criterion, the flow rules (Equation 15.8) with $\sigma_z = 0$ and $\rho = 0$ can be expressed as $0 = d\varepsilon_y/d\varepsilon_x = [\alpha^{a-1} + R(\alpha - 1)^{a-1}]/[R(1 - \alpha)^{a-1} + 1]$. $= [\alpha^7 - 1.75(1 - \alpha)^7]/[1.75(1 - \alpha)^7 + 1]$. Trial and error solution gives $\alpha = 0.520$.

Figure 15.4. Directional features in the microstructure of 2024-T6 aluminum sheet include grain boundaries, aligned and elongated inclusions. From *Metals Handbook, v.7, 8th Ed.* ASM (1972).

Anisotropy of Fracture

Mechanical working tends to produce directional microstructures, as shown in Figures 15.4 to 15.6. Grain boundaries and weak interfaces are aligned by the working. Inclusions are elongated and sometimes broken up into strings of smaller inclusions. Often loading in service is parallel to the direction along which the interfaces and inclusions are aligned, so the alignment has little effect on the ductility. For example, wires and rods are normally stressed parallel to their axes and the stresses in rolled plates are normally in the plane of the plate. In these cases, weak interfaces parallel to the wire or rod axis and inclusions parallel to the rolling plane are not very important. For this reason,

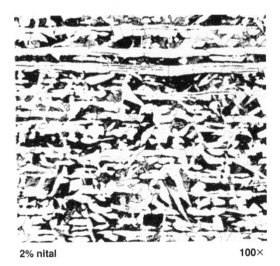

Figure 15.5. Microstructure of a steel plate consisting of bands of pearlite (dark regions) and ferrite (light regions). From *Metals Handbook*, v.7, 8th ed. ASM (1972).

2% nital   100×

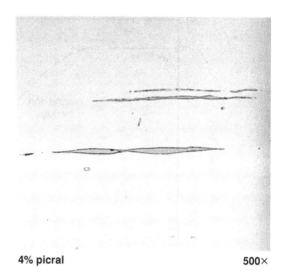

Figure 15.6. Greater magnification of the microstructure of the steel plate in Figure 15.4 showing elongated sulfide inclusions. These inclusions are the major cause of directionality of fracture in steel. From Metals Handbook, v. 7 8th ed. ASM (1972).

4% picral   **500×**

the anisotropy of fracture properties caused by mechanical fibering is often ignored.

Sometimes, the largest stresses are normal to the aligned fibers, and in these cases failures may occur by delamination parallel to the fiber axis. Welded T-joints of plates may fail this way by delamination. With severe bending, rods may splinter parallel to directions of prior working. Forged parts may fail along flow lines formed by the alignment of inclusions and weak interfaces. Annealing of worked parts does not remove such directionality. Even if recrystallization produces equiaxed grains, inclusion alignment is usually unaffected.

Anisotropy in Polymers

In linear polymers, the stresses required for yielding and crazing depend on the direction of applied tension. Crazing predominates if tension is applied perpendicular to the fiber alignment, as shown by Figure 15.7.

Notes

Rodney Hill (born 1921 in Leeds) earned his MA (1946), PhD (1948), and ScD (1959) from Cambridge University. He worked in the Armarment Research Department, Kent, Bitish Iron & Steel Research Association, Bristol University and Nottingham University before becoming a Professor of Mechanics of Solids at Cambridge University. His book, *Mathematical Theory of Plasticity* (Oxford University Press, 1950) is a classic. It contains original work on applications of slip line fields in addition to introducing the first complete theory of plastic anisotropy.

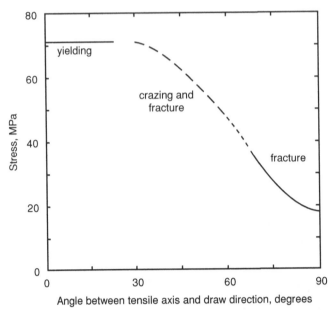

Figure 15.7. Effect of testing direction on the stresses necessary for yielding and for crazing at 20°C in polystyrene that has been oriented by stretching 260%. Data from L. Cramwell and D. Hull, *Int. Symp. on Macromolecules* (1977).

REFERENCES

R. Hill, *Mathematical Theory of Plasticity*, Oxford U. Press, 1950.
W. F. Hosford, *Mechanics of Single Crystals and Textured Polycrystals*, Oxford U. Press, 1992.
W. F. Hosford, *Mechanical Behavior of Materials*, Cambridge Univ. Press, 2005.

Problems

1. Calculate Young's modulus for an iron crystal when tension is applied along a <122> direction.

2. Zinc has the following elastic constants:

$$s_{11} = 0.84 \times 10^{-11} \text{ Pa}^{-1}; \quad s_{33} = 2.87 \times 10^{-11} \text{ Pa}^{-11}; \quad s_{12} = 0.11 \times 10^{-11} \text{ Pa}^{-1}$$
$$s_{13} = -0.78 \times 10^{-11} \text{ Pa}^{-1}; \quad s_{44} = 2.64 \times 10^{-11} \text{ Pa}^{-11}; \quad s_{66} = 2(s_{11} - s_{12})$$

   Find the bulk modulus of zinc.

3. Calculate the effective Young's modulus for a cubic crystal loaded in the [110] direction in terms of the constants, s_{11} s_{12}, and s_{44}. Do this by assuming uniaxial tension along [110] and expressing $\sigma_1, \sigma_2, \ldots \sigma_{12}$ in terms of $\sigma_{[110]}$. Then use the matrix of elastic constants to find $e_1, e_2, \ldots \gamma_{12}$, and finally resolve theses strains onto the [110] axis to find $e_{[110]}$.

4. When a polycrystal is elastically strained in tension, it is reasonable to assume that the strains in all grains are the same. Using this assumption,

calculate for iron the ratio of the stress in grains oriented with <111> parallel to the tensile axis to the average stress, $\sigma_{111}/\sigma_{av} = E_{111}/E_{av}$. Calculate σ_{100}/σ_{av}.

5. Using the 1948 Hill criterion for a sheet with planar isotropy,
   a) Derive an expression for $\alpha = \sigma_y/\sigma_x$ for plane strain ($\varepsilon_y = 0$) and plane stress ($\sigma_z = 0$).
   b) Find the stress for yielding under this form of loading in terms of X and R.
   c) For a material loaded such that $d\varepsilon_z = 0$ and $\sigma_z = 0$, calculate σ_x and σ_y at yielding.
   d) For a material loaded such that $d\varepsilon_x = d\varepsilon_y$ and $\sigma_z = 0$, calculate σ_x and σ_y at yielding.

6. Consider a sheet with planar isotropy (equal properties in all directions in the sheet) loaded under plane stress ($\sigma_z = 0$).
   a) Express the ratio, $\rho = \varepsilon_y/\varepsilon_x$, as a function of the stress ratio, $\alpha = \sigma_y/\sigma_x$.
   b) Write an expression for $\bar{\sigma}$ in terms of α, R, and σ_x.
   c) Write an expression for $d\bar{\varepsilon}$ in terms of α, R, and $d\varepsilon_x$. Remember that

$$\bar{\sigma}\ d\bar{\varepsilon} = \sigma_x d\varepsilon_x + \sigma_y d\varepsilon_y + \sigma_z d\varepsilon_z.$$

7. Take the cardboard back of a pad of paper and cut it into a square. Then support the cardboard horizontally with blocks at each end and apply a weight in the middle. Measure the deflection. Next, rotate the cardboard 90° and repeat the experiment using the same weight. By what factor do the two deflections differ? By what factor does the elastic modulus, E, vary with direction? Why was the cardboard used for the backing of the pad placed in the orientation that it was?

8. Consider the matrix of elastic constants for a composite consisting of an elastically soft matrix reinforced by a 90° cross-ply of stiff fibers. The general form of the matrix of elastic constants is

$$
\begin{matrix}
s_{11} & s_{12} & s_{13} & 0 & 0 & 0 \\
s_{12} & s_{11} & s_{13} & 0 & 0 & 0 \\
s_{13} & s_{13} & s_{33} & 0 & 0 & 0 \\
0 & 0 & 0 & 0 & s_{44} & 0 \\
0 & 0 & 0 & 0 & 0 & s_{66}
\end{matrix}
$$

Young's modulus for the composite when loaded parallel to one of the sets of fibers is 100 GPa, so $s_{11} = 10 \times 10^{-12}\,\mathrm{Pa}^{-1}$. Of the values listed here, which is most likely for s_{12}? For s_{66}? a. $10 \times 10^{-10}\,\mathrm{Pa}^{-1}$; b. 30×10^{-12} Pa^{-1}; c. $10 \times 10^{-12}\,\mathrm{Pa}^{-1}$; d. $3 \times 10^{-12}\,\mathrm{Pa}^{-1}$; e. $100 \times 10^{-12}\,\mathrm{Pa}^{-1}$; f. $-10 \times 10^{-10}\,\mathrm{Pa}^{-1}$; g. $-30 \times 10^{-12}\,\mathrm{Pa}^{-1}$; h. $-10 \times 10^{-12}\,\mathrm{Pa}^{-1}$; i. -3×10^{-12} Pa^{-1}; j. $-100 \times 10^{-12}\,\mathrm{Pa}^{-1}$.

Index

Printed in the United States
By Bookmasters